Crimes Against the Environment

Crimes Against the Environment explains the seriousness of the threat posed by pollution, its roots, how it has evolved, how it differs across the planet, and how society has endeavored to create and enforce laws directed at its control.

Rebovich and Curtis begin with an overview of hazardous waste, the industries that produce toxins, available methods of waste treatment, and the legal environment of environmental crime. They examine the forces driving criminal behavior and the methods offenders adopt, as well as protections against polluters and their effectiveness. The book concludes with an examination of environmental justice in the United States and globally, and looks ahead to the future of crime control and prevention in this arena. Case studies and discussion questions offer further perspective on these challenging issues of environmental integrity.

This text serves undergraduate or early-stage graduate students majoring in criminal justice, environmental science, sociology, and political science, and could also serve as a resource for professionals in environment-related occupations.

Donald J. Rebovich, Ph.D., Distinguished Professor of Criminal Justice at Utica College, Utica, N.Y., is the Director of Financial Crime Programs at the College. Dr. Rebovich is also the Executive Director of the Center for Identity Management and Information Protection (CIMIP) of Utica College. Prior to Utica College, Dr. Rebovich served as the Research Director for the National White Collar Crime Center (NW3C) and the National District Attorneys Association (NDAA). At NW3C he was responsible for directing the national analysis of Internet crime report data generated by the FBI's Internet Fraud Complaint Center, and directing the National Public Survey on White Collar Crime program. He is co-author of "Identity Fraud Trends and Patterns: Building a Data-Based Foundation for Proactive Enforcement," a national study of U.S. Secret Service identity theft cases covering a six-year period. He is also the author of *Dangerous Ground: The World of Hazardous Waste Crime*, which presented the results of the first empirical study of environmental crime and its control in the United States. Most recently, he co-edited a text entitled *The New Technology of Crime, Law and Social Control*. His background includes research in identity crime characteristics, economic crime victimization, white collar crime prosecution, protected health information, human trafficking, and multijurisdictional task force development.

George E. Curtis is Professor Emeritus of Criminal Justice at Utica College, Utica, N.Y. He previously served as the Dean of the School of Business and Justice Studies, Executive Director of the Economic Crime Institute of Utica College, and as Director of Economic Crime Programs at the College. He received a Bachelor of Arts degree from Syracuse University in 1964 and a Juris Doctor degree from Brooklyn Law School in 1967. He is an attorney admitted to the Bar of the State of New York. Curtis served as a confidential law clerk in the New York court system for more than 26 years and is a former President of his local bar association and a former delegate to the New York State Bar Association's House of Delegates. Professor Curtis is the author of *The Law of Cybercrimes and Their Investigation*, which was published in August 2011 by CRC Press, and is a co-author, with R. Bruce McBride, of a college text on Proactive Security Administration, published by Pearson Prentice Hall. Professor Curtis currently teaches undergraduate law-related courses in economic crime and cybercrime.

Crimes Against the Environment

Donald J. Rebovich & George E. Curtis

NEW YORK AND LONDON

First published 2021
by Routledge
52 Vanderbilt Avenue, New York, NY 10017

and by Routledge
2 Park Square, Milton Park, Abingdon, Oxon, OX14 4RN

Routledge is an imprint of the Taylor & Francis Group, an informa business

© 2021 Taylor & Francis

The right of Donald J. Rebovich & George E. Curtis to be identified as authors of this work has been asserted by them in accordance with sections 77 and 78 of the Copyright, Designs and Patents Act 1988.

All rights reserved. No part of this book may be reprinted or reproduced or utilised in any form or by any electronic, mechanical, or other means, now known or hereafter invented, including photocopying and recording, or in any information storage or retrieval system, without permission in writing from the publishers.

Trademark notice: Product or corporate names may be trademarks or registered trademarks, and are used only for identification and explanation without intent to infringe.

Library of Congress Cataloging-in-Publication Data
Names: Rebovich, Donald J., author. | Curtis, George E., 1942- author.
Title: Crimes against the environment/Donald J. Rebovich & George E. Curtis.
Description: New York, NY: Routledge, 2021. | Includes bibliographical references and index.
Identifiers: LCCN 2020020459 (print) | LCCN 2020020460 (ebook) | ISBN 9780367902759 (hbk) | ISBN 9781498754866 (pbk) | ISBN 9781315380179 (ebk)
Subjects: LCSH: Offenses against the environment. | Pollution prevention. | Hazardous waste management industry. | Hazardous wastes–Law and legislation.
Classification: LCC HV6401 .R44 2021 (print) | LCC HV6401 (ebook) | DDC 364.1/45–dc23
LC record available at https://lccn.loc.gov/2020020459
LC ebook record available at https://lccn.loc.gov/2020020460

ISBN: 978-0-367-90275-9 (hbk)
ISBN: 978-1-4987-5486-6 (pbk)
ISBN: 978-1-315-38017-9 (ebk)

Typeset in Bembo
by Deanta Global Publishing Services Chennai India

Visit the eResources: www.routledge.com/9781498754866

Contents

Introduction	vi
1 Hazardous Waste: What Is It and Why Is It a Threat?	1
2 What Industries Produce Hazardous Waste and What Wastes Do They Produce?	15
3 The Legitimate Treatment of Hazardous Waste in the United States: What Was Done in the Past and What Is Done Today	22
4 The Legal Environment of Environmental Crime	32
5 The Driving Forces of Environmental Criminality	44
6 Criminal Methods of Today's Average Environmental Offender	60
7 Who Protects Us from Environmental Crime and How Effective Are They?	81
8 Justice for All? Are We Achieving Environmental Justice in the United States?	89
9 Environmental Crime Around the World	101
10 Addressing Environmental Issues for the Future	115
Index	127

Introduction

Crime. When the average person thinks of the word, it is safe to say that images of threat and danger probably come to mind. But those images are most likely associated with "street crimes" committed through force or threat of force, crimes such as robbery, assault, or sexual abuse. These are crimes in which the threat is palpable. These are crimes in which the threat is seen and experienced overtly through direct contact with the offender and the victim. *Crimes Against the Environment* focuses on crimes that, on the whole, may not seem as "direct" but may be as dangerous, if not more dangerous. These are crimes that are insidious because they can seem, on the surface, to be innocuous, yet can be deadly to wide swaths of populations and to nature itself. This pollution has been with us since the dawn of mankind and grows throughout the world every day. This book is dedicated to explaining the seriousness of this threat, its roots, how it has evolved, how it differs across the planet, and how society has endeavored to create and enforce laws directed at its control.

The first part of the book describes what pollution is as well as its consequences. It takes us back in time to an age when civilization had little knowledge of the effects of exposure to toxic chemicals in air, land, and water. The early chapters acquaint the reader with the parameters of toxicity and how we have reached the state we are in today. The potential effects of exposure to hazardous waste are discussed, along with efforts to treat and contain these wastes. In terms of criminality, the book delves into the reasons that offenders turn to committing these crimes and the forces that drive them. In addition, the book examines the methods that offenders have used to commit their acts and how they have attempted to cover up any signs that they were guilty of their criminal acts. Recognizing that crimes against the environment are not just a national problem but a global one, the authors examine how these crimes may vary depending upon where in the world they take place.

Later portions of this book tackle the subject of what environmental threats and their control may look like in the future. As with any type of future projection, some of the information here can be considered speculative. However, these projections are grounded in what we know presently about the trends of pollution expansion, legal mechanisms that are being put into place to control environmental violations, waste treatment, and changes encouraged through the presidential administration taking place at the time of the publication of this book.

Several chapters focus on laws and regulations that are designed to protect the environment through the imposition of criminal sanctions; the state and federal administrative agencies that enforce those laws as well as non-governmental organizations (NGOs) that strive to ensure compliance with statutory and regulatory requirements; and efforts to ensure racial and socio-economic justice in environmental decision-making. Although the emphasis of this book is on environmental *crime*, an examination of the various environmental protections is essential to an understanding of the discussion concerning administrative enforcement and the imposition of criminal penalties. For example, to determine whether a business or individual has committed a criminal violation of the Clean Water Act, it is essential to understand at least the basic requirements of that Act. Our chapter on environmental crime law, therefore, considers a wide range of federal and state environmental laws that provide for civil and criminal penalties. Additionally, we consider an area that, though technically not always a crime, the effects clearly contravene public policy: the topic of environmental justice.

The book contains two features that we hope readers will find helpful in the study of environmental crime: case studies and chapter-ending discussion questions. The case studies are designed to illustrate important areas of the chapter, and the discussion questions seek to provoke further thought, research, and discussion of important issues in environmental crime law.

Chapter 1

Hazardous Waste: What Is It and Why Is It a Threat?

One of the greatest environmental threats to modern society is something we typically refer to as "hazardous waste." The term can be used very loosely by the media and in general parlance, but there are certain common misconceptions and uncertainty in what the term exactly represents. Obviously, the term itself can bring to mind images of disastrous, catastrophic accidents that send toxins into the air, water, or on the ground that can threaten all forms of life. But, in truth, such images only explain part of the entire picture. Forms of the release of hazardous wastes, either by accident or intentionally, can run a wide continuum, from minor incidents perpetrated by individuals (e.g., the release of polluted crankcase oil by the owner of a motor vehicle at the end of an oil change), to the owner of a dry-cleaning shop dumping process residue down a city sewer, to a Fortune 500 company routinely dumping in a remote geographic area. But what are these substances, and where do they come from?

In general, hazardous wastes can be defined as toxic byproducts of entities such as manufacturing industries, farming activities (e.g., improper or excessive application of pesticides), septic systems, construction, auto body shops, laboratories, hospitals, and other industries. In terms of form, hazardous wastes often take the form of a liquid, a solid, or a type of sludge. Whatever the form, they will, to some degree, contain harmful chemicals, heavy metals, radiation, or pathogens. The key difference between this category of waste and common garbage (or "solid waste") is that these wastes are certain to be harmful to plants, animals, and human life if there is direct and continual contact through disposed chemicals that seep into groundwater systems through ground disposal, direct disposal into waterways, or through air pollution. A most insidious dimension of improperly or illegally disposed hazardous wastes is that their toxicity can "accumulate," or build upon itself, such that wastes dumped onto ground surfaces can continue to present health dangers years or even decades after they are dumped. In addition, the toxins can travel through a "biological chain," linking disposal to animal life affected by the toxins and, ultimately, to humans who absorb the toxins through the digestion of the flesh of the animals infected (National Geographic, 1999).

The most common legal definition of hazardous waste is supplied through the Resource Conservation and Recovery Act of 1976. Through that Act, the U.S. Congress stated that hazardous waste is any refuse that is considered hazardous because of its quantity, concentration, or physical, chemical, or infectious characteristics. The Act goes on to state that the characteristics may cause or contribute to an increase in mortality or an increase in serious irreversible illness. In addition, the Act explains that such wastes can be characterized as hazardous if they pose a substantial present or potential hazard to human health or the environment if they are improperly treated, stored, transported, or disposed of. The Act is quite specific in that it designates eight variations of toxicity that must be proven in criminal prosecutions. The Act names three possible means for these wastes to cause the most harm to the environment and human health. Those three means are surface water contamination, groundwater contamination, and air pollution.

Hazardous wastes are incorporated into lists published by the U.S. Environmental Protection Agency (EPA). These lists are organized into three categories. The F-list (non-specific source wastes) identifies wastes from common manufacturing and industrial processes, such as solvents that have been used in cleaning or degreasing operations. Because the processes producing these wastes can occur in different sectors of industry, the F-listed wastes are known as wastes from non-specific sources. The K-list (source-specific wastes) includes certain wastes from specific industries, such as petroleum refining or pesticide manufacturing. Certain sludges and wastewaters from treatment and production processes in these industries are examples of source-specific wastes. The P-list and the U-list (discarded commercial chemical products) include specific commercial chemical products in an unused form. Some pesticides and some pharmaceutical products become hazardous waste when discarded. Taken together, these three categories of hazardous wastes encompass the whole of hazardous wastes as defined by the United States (U.S. Environmental Protection Agency, 2006).

More and more, the general public is becoming increasingly aware of the threats presented by hazardous waste pollution. This is often brought to the attention of the public by high-profile cases that involve either intentional or accidental pollution of the air, waterways, and land. Although much of the pollutants are born from technological advances in our society and the byproduct of those advances, pollution is far from a new problem. It has been with us for ages and will continue to be a problem vexing present society and societies of the future.

How Did We Get to This Point? A Chronology from Ancient History to the Present

Although hazardous pollution became a "hot button" topic in the 1970s and 1980s, especially since several high-profile violations captured public attention during that period, it hardly represented the early phases of such pollution. Some contend that the early uses of wood-burning fires proved to be some of the original sources of air pollution. Ancient Romans were responsible for some of the first examples of lead pollution through their active use of smelters. Ancient Romans were also responsible for one of the earliest known pollutions of a major river, the Tiber River. During the third century BC, Romans thought nothing of dumping human waste into the Tiber River via their sewers. Eventually, the Romans built aqueducts so that they would have ample, clean drinking water. While pollution at this time did not generally have wide-reaching effects on the environment, the evolution of humans moving from nomadic to settled societies led to pollution becoming more expansive and represented a serious threat to our ancestors. Pollution practices like those of the Ancient Romans would be replicated in other societies years later and become the source of typhoid and cholera outbreaks through the centuries in many different lands (Lofrano and Brown, 2010; Havlicek and Morcinek, 2016).

According to scholars of environmental history like Makra and Brimblecombe, lead mining originally began around 4000 BC. They point out that 1,000 years later, a new smelting technology had been developed to extract lead from sulfide ores of lead, resulting in an acceleration of lead mining and smelting. The golden age of the Roman Empire was responsible for a maximum lead production of over 80,000 tons a year. The Romans of that time came to realize the deleterious effects of lead by taking note of symptoms suffered by workers in lead mines. At that time, Romans began coating their bronze and copper cooking pots with lead, as well as in urns used to contain wine, to prevent leaching and preserve flavors. Some historians have speculated that lead poisoning could conceivably have been the cause of the unusual behavior patterns exhibited by Emperor Caligula and Emperor Nero. Evidenced by documentation of special environmental legislation designed to protect its unique environment, the ancient city of Jerusalem enacted specific laws regulating the location of tanning factories. Such factories were prohibited from being located less than 60 cubits from the city wall and were located to the eastern side of the city because of northerly and westerly winds that could blow pollution back into the city (Makra and Brimblecombe, 2004).

In the Far East, researchers have found evidence of metal pollution going back as far as 1500 BC. There, geochemical measurements on sediment cores from China's Lake Erhai were used to determine the timing of changes in metal concentrations over 4,500 years in Yunnan, a borderline region in southwestern China known for rich mineral deposits. Their findings uncovered the environmental legacy of human exploitation of natural resources in Yunnan. The analysis of sediment cores from the lake bed revealed metal pollutants like copper, lead, silver, cadmium, and zinc. The researchers noticed an increase in concentrations of these heavy metals between AD 1100 and 1300 and concluded that they are associated with an upsurge in silver smelting during that period. Culminating during the rule of the Mongols between AD 1271 and 1368, the researchers found that these concentrations approached levels three to four times higher than those from industrialized mining activity occurring today. In their report, the researchers underscore that the concentrations of metal from the smelting practices are harmful to aquatics organisms and contribute to known issues in modern-day sediment quality (Hillman, Abbott, Yu, Bain and Chiou-Peng, 2015).

In 14th-century Europe, city administrators found it practically impossible to control and prevent water pollution problems. In 1307, London officials concluded that tanning wastes were the source of water pollution in the fleet river. This pollution, combined with the problems emanating from London's blocked sewers, created teeming cesspools affecting citizens' wells. London officials concluded it prudent to proclaim that air pollution from the burning of coal could meet strict fines. The 1830s ushered in a lingering outbreak of cholera in England. No one imagined that pollution could be the source. This lack of awareness was found to be, quite sadly, common to the existence of pollution throughout history, a lack of understanding of or the outright refusal to accept the "cause and effect" principle. Curiously, many "authorities" in the London of the 1830s attributed that city's devastating cholera outbreak in 1832 to a bad smell in the air termed "miasma." One of the supporters of this theory at the time happened to be Florence Nightingale, which lent the theory an aura of credibility considering Ms. Nightingale's irreproachable reputation. Were it not for the unwavering diligence of the physician and father of modern epidemiology, Dr. John Snow, the association between bacteria in dumped sewage and cholera would not have been made. As it was, the illogical belief in the miasma explanation proved so strong that Dr. Snow's theory was not accepted until over 20 years after the original outbreak and, ultimately, after Dr. Snow's death. Despite irrefutable evidence submitted to London's health agency, the St. James Parish Vestry, it took a sharp escalation of pollution and cholera to finally convince politicians to act. In 1858, the state of the River Thames was described as "The Great Stink," and public outrage compelled leaders to come up with an effective response to the horrific condition of the Thames. The creation of an innovative sewer system by Joseph Bazalgette led to the cleansing of the Thames of bacteria responsible for waterborne diseases and a significant decrease in incidents of cholera (Westminster Archives, 2015).

By the mid-1800s, England's Royal Commission on River Pollution described water that was "poisoned, corrupted, and clogged" from refuse from mines, chemical works, and dyeing processes. During that same time period in nearby Wales, the town of Llanelli dealt with its own special problem dealing with noxious fumes coming from the towns in its copper works by building the world's highest stack at that point in time (320 feet). Through the 1880s, copper industries in Wales gave way to lead, nickel, and silver industries presenting new pollution problems for that part of the world. Farmers in Wales increasingly were reporting rising numbers of cattle deaths with speculation of direct connections to the proliferation of the growing metal industries (Markham, 1994).

The Industrial Revolution saw the rapid increase in large-scale coal use during the latter part of the 19th century and the beginning of the 20th century. The new pervasive health threat became an abundance of soot and smog enveloping urban areas in Europe and the United States, damaging the health of citizens living in and around these areas. Moving forward into the 1940s, "industrial smog" was reported as being the cause of the 1948 asphyxiation deaths of 20 people in Donora, Pennsylvania, as well as the sickening of thousands in Donora's suburbs. The reaction to this incident was strong. It led to the creation of the Air Pollution Control Act of 1955 and represents the first federal attempt to control air pollution in the United States. On the heels of this tragedy came the deaths of over 4,000 citizens of London in 1952, deaths attributed to industrial air pollution. The same type of coal-powered plants was found, at that time, to be the source of surging levels of "acid rain". Eventually, coal-burning pollution gave way to pollutants from motor vehicles as a primary source of air pollution (Ferrara, 2015; Snyder, 2008.) The use of leaded gasoline raised lead levels in populations around the world. Leaded gasoline was phased out in the United States starting in 1976 but is still in use in many parts of the world.

Today's Waste Problems

Today's hazardous waste problems in the United States can be said to originate as an offshoot of America's technological explosion during the late 1940s and early 1950s. Starting in this period, the pace of manufacturing new chemicals rose rapidly and included the production of chemicals that were primary ingredients for pesticides, plastics, and detergents. The incidence of some present human cancers has been typically linked to the latent effects of exposure to illicitly disposed chemicals of post-World War II. After World War II, petroleum became the key ingredient in the manufacture of synthetic organic chemicals. This became a game changer for the disposal of hazardous wastes. For organic chemicals, the common element of carbon links with itself to create strings of organic chemical compounds. Synthetic organic chemicals mimic the characteristics of organic chemicals, paving the way for the creation of plastics, electronic components, and construction materials. Along with these materials came the introduction of polychlorinated biphenyls (PCBs) and inorganic pesticides like DDT. These are substances that are not biodegradable, accumulate in the environment, and had a hand in increasing rates of cancer, birth defects, and mental retardation. In the United States, the waterways in the Northeast have been the recipient of many industrial chemicals and wastes, including sulfuric acid, soda ash, muriatic acid, limes, and dyes from industrial mills (Markham, 1994).

It is uncertain whether inappropriate waste disposal since the post-World War II period can be largely attributed to industry ignorance or to industry irresponsibility. Before the advent of the implementation of waste reduction policies by waste-producing industries, the 1970s and 1980s witnessed some of the highest waste-producing volumes in the United States. Some alleged a combination of corporate naïveté regarding environmental damage, corruption, and regulatory inefficiencies as the chief causal factors (Epstein, Brown and Pope, 1982). Others are less charitable to generating companies. Wolf (1983) asserted that industry would often attempt to exonerate itself from blame by rationalizing that in the past, generators had been ignorant of the possible damage caused by their disposal methods and did not expect that contracted disposal firms would dump them improperly. Wolf questioned these explanations and concluded that industry had consciously disregarded potential harmful consequences in an unconscionable quest of reduced disposal costs.

Brownstein (1981) added that corporate quests to expand production and to discover new markets exacerbated the problem of the growing volume of waste products starting in the 1980s. The economies of scale were viewed as considerable, especially for the petroleum industry, due to its dependence on expensive equipment. This factor added further pressure to cultivate new markets, which inevitably led to a generation of new waste materials and the consequential compromising of ethics and the disposal of these wastes. According to Brownstein, the chemical industry had been shirking the responsibility for proper disposal of wastes by claiming in the final analysis that society should bear the bulk of disposal responsibility since they benefit from chemical products.

Major Pollution Events Since the Mid-1900s

As time marched on, a number of major pollution events occurring in the United States and elsewhere put pollution and the threat presented to the environment into the spotlight. The following is a summary of these events.

Cuyahoga River

An unusual phenomenon occurred in the United States between the mid-1930s and the late 1960s in the form of fires erupting on the Cuyahoga River in Cleveland, Ohio. This river, which flows into Lake Erie, became so polluted that a spark from a blowtorch in 1936 set fire to floating oils and debris. Several additional fires occurred over the next 30 years culminating in a horrific fire that led to the creation of major legislation and federal agency development. The fire that started in 1969 captured national media coverage and was instrumental as an incentive to creating the Federal Water Pollution Control Act of 1972. This Act is also known as the Clean Water Act. The Act allows for the funding to improve sewage treatment plants (STPs) and sets restrictions on what can be discharged into water. The fires also were largely responsible for the impetus leading to the creation of the Environmental Protection Agency and the passage of the Oil Pollution Act of 1990. This Act prohibits the discharge of oil into navigable waters. Around that same time in the United States, the Clean Air Act was enacted, setting limits on the discharge of air pollutants from motor vehicles and industrial facilities (National Ocean and Atmospheric Administration, 2008).

Love Canal and Chemical Control Corporation

One such event was the Love Canal disaster. Originally, Love Canal was built to supply water power in the Niagara Falls, New York area. The project ran into problems, was abandoned, and consequently, the canal was turned into a landfill. The landfill was operated by the Hooker Chemical Company, which eventually sold the site to the Niagara Falls Board of Education. Houses and an elementary school sprouted on top of the site. After a record rainfall in 1977, wastes from the landfill leached into private property and into elementary school playgrounds. Severe illnesses and miscarriages were reported as being the result of human exposure to the leaching contaminants. Eventually, over 250 families in the surrounding area had to abandon their homes because of the hazardous waste contamination. The Love Canal experience drew attention to the longevity of the chemical toxicity of hazardous wastes, given the fact that the wastes were disposed of and buried decades before the fluids, and the danger associated with them, rose to the surface (Ferrara, 2008).

Shortly after the Love Canal disaster, another similar event occurred in nearby New Jersey that, in some ways, was even more dramatic. It involved the Chemical Control Corporation (CCC). The CCC site consisted of a two-acre parcel of land adjacent to the Elizabeth River. From 1970 to 1978, CCC operated as a hazardous waste storage, treatment, and disposal facility, accepting various types of chemicals including acids, arsenic, bases, cyanides, flammable solvents, polychlorinated biphenyls (PCBs), compressed gases, biological agents, and pesticides. Throughout its operations, CCC was cited for discharge and waste storage violations. In 1979, the state of New Jersey initiated a site cleanup that included removal of bulk solids and liquids, drums at and below the soil surface, gas cylinders, infectious wastes, radioactive wastes, highly explosive liquids, debris, tanks, and three feet of soil. Excavated soil areas were replaced with a three-foot gravel cover (Stranahan and King, 2000).

An explosion and fire in 1980 interrupted the CCC site cleanup and created additional cleanup needs. At 10:54 PM, April 21, 1980, CCC exploded. A mushroom cloud ripped into the sky above Elizabeth, New Jersey, and 55-gallon drums of solvents, pesticides, mercury compounds, explosives, and acids rocketed 200 feet into the air; 35,000 drums burned furiously. Contaminated smoke reigned over a 15-mile square area, including parts of New York City. The New Jersey National Guard was put on alert, and plans were made to evacuate over one million people from the area. After ten hours, the fire was under control. It could've been worse, had not state officials taken over the site a year earlier using a court order. What officials found there prompted them to call in military bomb squads to remove TNT, picric acid, and mustard gas. At the time, an official from the federal Bureau of Alcohol, Tobacco, and Firearms stated that the material found at CCC could have, at a minimum, caused a lethal dose to all of Staten Island and lower Manhattan in the event of another fire. After the fire and explosion, the preliminary cleanup was accelerated and was completed in 1981. The final cleanup took many more years (Stranahan and King, 2000).

The Love Canal and CCC events combined to force the hand of the U.S. government to take further legislative action to protect the public from dangerous pollutants. Congress passed the U.S. Resource Conservation and Recovery Act (RCRA). This Act was important in compelling the U.S. Environmental Protection Agency (EPA) to devise a cradle-to-grave national manifest system to track hazardous wastes and to develop a system of penalties for violations. In addition, an emergency fund was put in place to clean hazardous waste dumps either in cases where the owners of the dumps were unable to pay for cleanup or in cases where the owners were unknown. This legislation was known as the Comprehensive Environmental Response Compensation and Liability Act (CERCLA). This came to be commonly known as "Superfund" and is in effect up to the present. While this legislation was a step in the right direction, it had no effect on the next major environmental tragedy in the United States. This took place in Times Beach, Missouri, where the streets had been contaminated by a waste hauler contracted to spread oil on streets to tamp down dust. In this case, the federal government was compelled to relocate over 2,500 residents. The federal buy-out of the town totaled over $36 million of taxpayer funds (Schmidt, 1983).

Ciba-Geigy and R.W. Grace

New Jersey continued to be an environmental battleground with the Ciba-Geigy (a Swiss pharmaceutical firm) hazardous waste dumping problems starting in 1982. That year, their plant in Toms River was added to the EPA list of Superfund cleanup sites when more than 120 different chemicals were discovered in local groundwater. Two years later, a leak sprung in the plant's ten-mile long pipeline that ran underground stretching to the Atlantic Ocean, a pipeline that the residents at the Jersey shore area did not know about. The outcry of grassroots environmental groups and the shore public in New Jersey proved too much for the firm. Ciba shut down the pipeline and closed most of its buildings on the Toms River site. Eventually, Ciba relocated much of their U.S. operations to Louisiana and Alabama. In 1992, Ciba-Geigy pled guilty to illegal waste disposal and paid more than $60 million in fines and landfill and groundwater cleanup costs (Miller, 2013).

Jan Schlichtmann, an attorney who represented citizen groups in the Ciba-Geigy case, was a key player in an earlier case in Woburn, Massachusetts. This case focused on the alleged contamination of two municipal supply wells in Woburn by three local industries. Eight families whose residences were in close proximity to the wells were the plaintiffs against W.R. Grace & Co. Grace owned the Cryovac Plant, UniFirst Corporation, Interstate Uniform Services, and Beatrice Foods, Inc., owner of the John Riley Tannery. The allegations were that hazardous chemicals discovered in water samples from the wells were the cause of health problems experienced by local residents, including cases of leukemia, resulting in several deaths. Children of seven of the plaintiffs contracted leukemia. The suit was filed in May 1982. High concentrations of TCE were subsequently found in wells at the plant. Over a two-month period, the jury heard technical testimony from expert witnesses representing all parties in the case, and then they visited the wells and the proximity of the plaintiffs' and defendants' properties. After this examination by the jurors, W.R. Grace was found liable, and Beatrice Foods was found not liable for contaminating the wells. The plaintiffs' lawsuit against W.R. Grace & Co. moved forward, and the judge accepted a motion for a mistrial from Grace. W.R. Grace and the plaintiffs reached an $8 million settlement before the judge's ruling was made (Bair, 2008).

Union Carbide and Bhopal

A couple of years later, in 1984, India was the site of an environmental catastrophe of epic proportions: the Bhopal tragedy caused by the Union Carbide chemical company. On December

2, 1984, thousands of people in the Bhopal area died when over 30 times of methyl isocyanate gas was released from the Union Carbide plant. Further deaths and physical ailments persisted around the geographic area for years following the Union Carbide release. In 1985, Union Carbide released toxic gas in one of their plants in West Virginia, leading to serious illnesses affecting residents around the plant. Once again, U.S. Congressional legislation reacted by quickly enacting the Emergency Planning and Community Right to Know Act (EPCRA). This law compels any firm that, in any way, handles hazardous waste to present a full and complete disclosure on several critical pollution-related topics. It includes information on the state of storage and handling facilities and any accidental release of hazardous material in a quantity above an established safe limit. The EPA has reported a significant decrease in toxic releases since the enactment of this law (Ferrara, 2008).

Exxon Valdez, BP Oil, and Duke Energy

Five years after the Bhopal disaster, another environmental disaster occurred off the coast of Alaska. This was an accidental oil spill of the Exxon Valdez tanker resulting from the tanker running aground at Bligh Reef, Alaska. Eleven thousand gallons of oil spilled into Prince William Sound as a result of this accident. It was considered to be, at the time, the worst such oil spill in terms of damage to the environment. This particular event led to the enactment of the 1990 Oil Pollution Act. This law made it mandatory that oil tankers would be double-hulled as a prevention mechanism to prevent spills resulting from tanker accidents (Ferrara, 2008).

Another major oil spill occurred on April 20, 2010, in the Gulf of Mexico. On that day, the Deepwater Horizon oil rig operated by BP Oil exploded and sunk, killing 11 workers on the oil rig. Until it was capped on July 15, 2010, oil flowed from a seafloor oil gusher for 87 days as a result of the accident. Despite insisting that it was not chiefly responsible for the accident, BP Oil was ultimately held responsible by a federal judge in New Orleans. The judgment stated that BP Oil had rejected known risks and that the oil company was grossly negligent. Ultimately, BP Oil agreed to pay $18.7 billion in fines representing the largest corporate settlement in U.S. history (Robertson and Krauss, 2014).

A more recent example of an accidental discharge of pollutants occurred in North Carolina in February of 2015. Electric power Co. Duke Energy experienced a rupture in a 48-inch concrete, underground storm pipe that discharged toxic control coal ash slurry into North Carolina's Dan River. The coal ash proved to have high concentrations of selenium, mercury, and arsenic. Like the BP Oil incident, it took some time to plug the leak. The leak was plugged after eight days, but not until more than 30 million gallons of the coal ash was released into the Dan River, flowing more than 70 miles from the source of the rupture into the state of Virginia (Figure 1.1) (Katz, 2015).

Emerging Problems

The 21st century has presented formidable new challenges for the control and proper handling of hazardous wastes. For a society increasingly dependent on the use of computers and mobile devices, our modern world must contend with the ramifications of discarded electronic equipment associated with such computerized devices. Now known as E-waste, the wanton and proper

Figure 1.1 Aerial photo of Duke Energy coal ash spill. www.epa.gov/enforcement/2015-major-criminal-cases

disposal of these wastes has created a new health problem for the public. This is particularly so for Third World countries, which often become the final resting place for E-waste. Globally, this problem is considered so serious that it has become a priority for the United Nations and Interpol. There is a growing market for E-waste, especially in Africa, due to the fact that the costs associated with buying new electronic equipment are seen as being prohibitive. This creates a large market for used electronic materials. In many instances, it is found that the used electronic equipment is irrepairable, resulting in the indiscriminate dumping of these wastes.

A particular E-waste problem has been the illegal trade and disposal of cathode ray tubes. A cathode ray tube (CRT) is the glass video display component of an electronic device (usually a computer or television monitor). CRT funnel glass generally contains high enough concentrations of lead that the glass is regulated as hazardous waste when disposed. In the United States, efforts have been made to reign in illegal disposal and trade of CRTs. On June 26, 2014, the U.S. EPA published a final rule that revised the export provisions of the CRT final rule under the Resource Conservation and Recovery Act (RCRA) (79 FR 36220). The final rule became effective on December 26, 2014. These changes allow the EPA to obtain additional information to better track exports of CRTs for reuse and recycling in order to ensure the safe management of these materials. The changes were recommended by the Interagency Task Force on Electronics Stewardship, which consists of 16 federal agencies, in its report titled National Strategy for Electronics Stewardship (July 20, 2011). The final rule fulfills an EPA commitment toward the Task Force's goal to reduce harm from U.S. exports of E-waste and improve the safe handling of used electronics in developing countries. Specifically, the rule adds a definition of "CRT exporter" to eliminate potential confusion over who is responsible for fulfilling CRT exporter duties, including submitting the export notices. The rule also requires information on all interim and final destinations for CRTs exported for recycling to provide complete information to receiving countries (U.S. Environmental Protection Agency, 2014).

Another new-age hazardous waste problem is surfacing in the "fracking" industry. This is a term used to describe a method to reach sources of natural gas or petroleum. The fracking term refers to fracturing rock through hydraulic means. This method uses hydraulically pressured liquid, resulting in chemically toxic wastewater that must be properly treated to prevent threats to the public health of those living in proximity to fracturing operations. To avoid the cost of properly treating the substances, fracking operatives will dump the wastes in the surrounding surface area or in nearby waterways. Such was the case in 2014 in Pennsylvania in which more than 50,000 gallons of fracking wastewater was dumped into a local waterway (Fair, 2014).

Discussion

To this day, there are some who still believe that wide-scale environmental pollution is a fairly recent phenomenon. The first chapter illustrates that this is, by far, not the case. Evidence of pollution in ancient time periods points to eras as far back as 1500 BC in China and the Golden Age of Ancient Rome, where early metal smelting took its toll on the health of the general public. One can trace through subsequent historical periods and note the changing patterns and trends in the type and volume of chemical and organic pollutants that have, in one way or another, been released into the environment, threatening the health of scores of people during those times. Much of the form of these pollutants has been determined by advances in technology, manufacturing processes, and the public's demand for certain goods and products. The Industrial Revolution became a major event in these changes in that the rapid increase of large-scale coal used created a litany of health problems during the early stages of the 20th century.

A new world of environmental issues and problems emanated out of the post-World War II era and the manufacture of synthetic organic chemicals, introducing the environment to harmful elements like DDT and PCBs, and exacerbating the growing threat to public health. Moving through the 20th century, major environmental disasters occurred, some by accident, some intentional, all combining to result in a new age in which the public became frightfully aware of the potential consequences of handling and mishandling of dangerous substances created as byproducts of societal and technological advances.

We are presently at a time in which technological inroads are moving so quickly that there is no reason to believe that the dangers presented by new pollutants will not let up. Indeed, the ubiquitous presence of burgeoning volumes of E-waste and questions revolving around hydraulic fracturing processes present new and perplexing environmental protection issues. It remains to be seen what, in this context, the next challenges will be for the effective protection of public health.

Discussion Questions

1) This chapter takes the reader through several eras of history and follows how the threat of hazardous substances to the environment has evolved over time. Select two different time periods described in the chapter. How and why did the specific threats change over time? To what degree do you believe the public was aware or unaware of these threats? Explain your rationale for your answer to this question.

2) In the subsection that addresses major pollution events since the mid-1900s, compare the events described. In what ways do they differ? What common features do they possess? How could these events have been prevented?
3) Based upon this chapter and present forces that we must contend with now, how do you see environmental challenges for the future?

Case Study: Chemical Control Corporation and Its Criminal Brethren

In the 1970s, Newark Bay and the Arthur Kill (a narrow body of water between Elizabeth, New Jersey, and Staten Island, New York) were hotbeds for the illegal disposal of hazardous wastes. As briefly mentioned in this chapter, Chemical Control Inc. was the site of one of the most notorious hazardous waste disasters of that time period. In their book, *Hazardous Waste in America*, authors Samuel Epstein, Lester Brown, and Carl Pope depicted the details of what led up to this event. The Chemical Control facility was based in Elizabeth, New Jersey, on a 3.4-acre tract of land. In 1979, state inspectors found that the facility was housing over 40,000 55-gallon drums of highly toxic industrial waste, some of which were in rusted and leaking drums. It was reported by inspectors that some areas of the facility had drums stacked five high. One of the alarming discoveries made by investigators was that about 100 pounds of explosive picric acid and radioactive waste was found in a loft in the incinerator building. In addition, cylinders of mustard gas, a highly toxic nerve gas, were found on-site. Rusted and leaking drums were also found to be packed into one room labeled as the "pesticide room" and another labeled the "boiler room" (Epstein, Brown and Pope, 1982).

At the time, it was determined that the site posed an imminent risk for fire and explosion. The site was located about 500 feet from a high-pressure gas line and about 2,000 feet from a large liquid natural gas tank. It was projected that an explosion at the Chemical Control site would set off flammable materials nearby and that potentially thousands of people could be killed within minutes. The "doomsday" prediction was that a fire at the site would give rise to thick clouds of toxic smoke over the city of Elizabeth and that the clouds would drift to Staten Island and possibly even to Manhattan. Being unsuccessful in getting the company to clean up the operation, New Jersey State government took over the site (Epstein, Brown and Pope, 1982).

The Chemical Control site caught fire on the morning of April 21, 1980. Fifty-five-gallon drums exploded and shot into the air. At first, some nearby residents thought that the sounds of the explosions were thunder. Others in closer vicinity of the site likened what they saw to visions of an atomic blast. The toxic smoke clouds that were predicted became a reality, forcing the closure of schools in the area. However, the worst scenario was avoided due to the direction of the winds away from the most populated areas. Over 25 firemen were injured in the 10 hours it took to put out the fire. After cleanup, a second on-site fire took place in June of that year involving truckloads of chemical wastes (Epstein, Brown and Pope, 1982).

The history of what led up to the fire and explosion at Chemical Control is a case study of how a hazardous waste treatment firm can start out with honorable intentions and devolve into a criminal operation presenting serious health risks to those working at the operation and residents living near it. The owner of Chemical Control was an individual named William Carracino. He started out working at New Jersey's only authorized dump for plastic and chemical wastes. The primarily solid waste disposal facility was named Kin-Buc and was located in Edison in central New Jersey. This facility had to close in the mid-1970s after it was found that large amounts of chemical liquid wastes had been disposed there improperly, presenting a threat to public health. Contending that he left Kin-Buc because he was pressured by management to participate in illegal activities, Carracino purchased a barrel factory in Elizabeth, New Jersey's waterfront, and started Chemical Control Inc. Things went well for a while until Carracino decided to install an incinerator and attempted to raise money for that purpose. He eventually connected with Michael Collington, the president of Northeast Pollution Control, a small manufacturing company. In short order, Northeast Pollution acquired over 80% of Chemical Control stock, reducing Carracino and his partners to minority shareholders. Financial difficulties led to Chemical Control filing for bankruptcy, and it consequently was taken over completely by Northeast Pollution. Carracino remained as the plant manager and the incinerator was installed, leading to a sharp upsurge in business for Chemical Control, making it one of New Jersey's largest waste disposal facilities. Chemical Control became the "go-to" company responsible for incinerating and processing high volumes of flammable and hazardous chemicals. At the time, the U.S. Food and Drug Administration actually regarded the operation as being a "model" for effectively and legitimately recycling wastes (Epstein, Brown and Pope, 1982).

So how did Chemical Control go bad? There is some debate about that. Carracino pointed to Northeast Pollution Control as the culprit, accusing leaders in the firm of pressuring him to stockpile more inventory than he could reasonably process. He maintained that the parent company head, Collington, had actually at one time berated Carracino for putting out a small fire at the Chemical Control site instead of letting it burn and collecting insurance on damage to the facility. Over the course of a couple of years, the relationship between Carracino and Collington worsened as Northeast Pollution gained even greater domination over Chemical Control. Carracino eventually claimed that one altercation between the two led to Collington pulling a gun on Carracino and proclaiming that Northeast Pollution Control would assume full management of Chemical Control (Epstein, Brown, and Pope, 1982).

Shortly after that altercation, Carracino was indicted for the illegal disposal of hazardous waste, convicted of three counts of illegal dumping, and in 1978 was sent to prison to serve a two-year prison sentence. Ultimately, Carracino characterized himself as blameless in the disastrous condition of the Chemical Control facility before the major fire erupted in April of 1980, asserting that he was running a "clean operation" while he was plant manager. However, he also publicly stressed that the biggest problem standing in the way of proper hazardous waste treatment and disposal in New Jersey at the time was that the state lacked a sufficient number of approved waste disposal facilities to handle all of the state's hazardous wastes. He lamented that in many cases, honest recyclers in New Jersey had no place to dispose of their hazardous wastes (Epstein, Brown and Pope, 1982).

But this story doesn't end here. It gradually became clear that criminality at Chemical Control was not an aberration for treatment and disposal facilities in Central and Northeastern New Jersey. The tentacles of criminality spread further than Chemical Control. Further investigations found that Northeast Pollution Control also owned another illegal operation besides Chemical Control. This one was called the A to Z Chemical Company, and it was located in New Brunswick, New Jersey, not far from the original Kin-Buc site. At the time, New Jersey State government officials characterized A to Z as being cut from the "same cloth" as Chemical Control. In time, state authorities found that Chemical Control had stored over 500 55-gallon drums of waste at the Duane Marine Salvage Corporation, which was another firm that was under an ongoing investigation for suspected environmental violations. Duane Marine was located in Perth Amboy on the aforementioned Arthur Kill waterway separating that part of New Jersey from Tottenville, Staten Island. The owners of Duane Marine were indicted in September of 1984 for illegally disposing of an estimated 500,000 gallons of hazardous wastes into Perth Amboy's sewer system. Back in the vicinity of the original Chemical Control site in Elizabeth, NJ, a company called Iron Oxide was revealed to have purportedly constructed an acid-waste treatment facility. There was only one problem – it didn't exist. State law enforcement personnel eventually conducted a raid on the site and found drainage pipes with external hookups to tanker trailers. With the help of the U.S. Army Corps of Engineers, investigators were able to trace the pipe under the site's building and directly into the Arthur Kill where over 50,000 gallons of toxic acidic chemical wastes were dumped (Epstein, Brown and Pope, 1982).

Such was the "Chemical Control-type" environmental crime terrain in New Jersey in the 1970s and 1980s. Times have changed, and the aggressive crime enforcement operations that were developed in the 1990s and into the 21st century are largely responsible for dramatically reduced environmental crime in that state. Hopefully, New Jersey will never return to that earlier period in which environmental crime was rampant.

References

Bair, S. (2008). The Woburn toxic trial. *Science in the Courtroom*. Retrieved from http://serc.carleton.edu/woburn/Case_summary.html

Brownstein, R. (1981). The toxic tragedy. In R. Hader, R. Brownstein, & J. Richard (Eds.), *Who's poisoning America: Corporate polluters and their victims in the chemical age*. San Francisco, CA: Sierra Club Books.

Epstein, S., Brown, L., & Pope, C. (1982). *Hazardous waste in America*. San Francisco, CA: Sierra Club Books.

Fair, M. (2014, June 6). Pennsylvania AG charges fracking waste hauler for illegal dumping. *Port Folio Media*. Retrieved from http://www.law360.com/articles/545474/pa-ag-charges-fracking-waste-hauler-for-illegal-dumping

Ferrara, A. (2008). *Pollution issues*. Retrieved from http://www.pollutionissues.com/Fo-Hi/History.html

Hillman, A., Abbott, M., Yu, J., Bain, D., & Chiou-Peng, T. (2015). Environmental legacy of copper metallurgy and Mongol silver smelting recorded in Yunnan Lake sediments. *Environmental Science and Technology, 49*(6), 3349–3357.

Katz, J. (2015, February 20). Duke Energy is charged in huge coal ash leak. *The New York Times*. Retrieved from http://www.nytimes.com/2015/02/21/us/duke-energy-is-charged-in-huge-coal-ash-leak.html

Makra, L., & Brimblecomb, P. (2004). Selections from the history of environmental pollution, with special attention to air pollution. Part 1. *International Journal of Environment and Pollution, 22*(6), 641–656.

Markham, A. (1994). *A brief history of pollution*. New York: St. Martin's Press.

Miller, P. (2013, May 7). Reflections on "chemical town." *Berkeley Patch*. Retrieved from http://patch.com/new-jersey/berkeley-nj/ciba

National Geographic. (1999). Toxic waste: Man's poisonous byproducts. Here's what you need to know about the warming planet, how it's affecting us, and what's at stake, Washington DC.

Robertson, C., & Krauss (2014, September 4). BP may be fined up to $18 billion for spill in gulf. *The New York Times*. Retrieved from http://www.nytimes.com/2014/09/05/business/bp-negligent-in-2010-oil-spill-us-judge-rules.html?_r=0

Schmidt, W. (1983, February 25). Denver lawyer's role in EPA decisions is focus of inquiries by Congress. *The New York Times*, p. 12.

Snyder, L. (2008). Donora, Pennsylvania. *Pollution Issues*. Retrieved from http://www.pollutionissues.com/Co-Ea/Donora-Pennsylvania.html

Stranahan, S., & King, L. (2000, April 30). Beyond the flame: Chemical fires brought litany of health problems. *Philadelphia Inquirer*, p. 9.

United States Environmental Protection Agency. (2006, June). *Wastes – Hazardous wastes*. Washington, DC: United States Environmental Protection Agency.

United States Environmental Protection Agency. (2014). *Fact sheet: 2014 revisions to the export provision of the cathode ray tube (CRT) final rule*. Retrieved from http://www3.epa.gov/epawaste/hazard/recycling/electron/crt_rul_fs_070114.pdf

Westminster Archives. (2015). The fascinating story of London's battle against cholera. *Cholera and the Thames*. Retrieved from http://www.choleraandthethames.co.uk/

Wolf, S. (1983). Hazardous waste trials and tribulations. *Environmental Law, 13*(2), 367–428.

Chapter 2

What Industries Produce Hazardous Waste and What Wastes Do They Produce?

As noted in Chapter 1, hazardous wastes are typically toxic byproducts of process activities integral to entities in a variety of areas, large and small. This can include manufacturing processes employed by large, mid-sized, and small businesses. It can also include runoff from farming operations and building construction sites. Medical laboratories and hospitals also produce chemical and infectious wastes that can be harmful to the environment. Auto body shops represent a primary source of chemical wastes as a residue of auto repair and painting activities. In addition, take into account the end users. These include homeowners who dispose of household chemical cleaners and disinfectants improperly. Taken together, they all combine to generate an enormous volume of hazardous waste that must be properly treated and disposed of so as not to present a direct threat to the environment and to public health.

Throughout history, advances in technology have resulted in the growth of special types of hazardous wastes known as byproducts. As mentioned earlier, a milestone in this pattern was the point at which petroleum began to be used as a primary ingredient in the manufacture of synthetic organic chemicals. Up until World War II, the world had to contend with wastes that were made up of organic chemicals. After World War II, scientific advances allowed for the imitation of the wide spectrum of characteristics of organic chemicals. This was done for the creation of synthetic organic chemicals that became key for producing certain electronic components, construction materials, and, most important, plastics. Consequently, the volume of hazardous wastes increased exponentially, creating a public burden for the proper treatment and disposal of these wastes.

A Special Case: Plastics

Since World War II, in the United States as in other countries, the plastics industry has increasingly become a part of our lives. In two instances, the new role of the arrival of plastics has been transmitted through American cinema. In the famous 1947 film *It's a Wonderful Life*, the protagonist, George Bailey, is offered an opportunity to get in on the ground floor of an industry producing plane parts of lightweight, heavy-duty plastic. Three decades later, the main character from the 1967 film *The Graduate*, Benjamin Braddock, is approached by a friend of his parents at his graduation party. The aimless Braddock confides to Mr. Maguire that he is unsure about his future. In the famous scene, Mr. Maguire whispers into his ear the word "Plastics." No explanation is given to Benjamin, leaving him confused. But most audiences got the inside joke; plastics have become a powerful presence in our society, and have come to represent a fixed feature in contemporary life and, to some, artificiality. Plastics had become a modern societal centerpiece, representing many things to many different people.

At a quick pace, new varieties of plastics were introduced into our society, plastics like polyurethane, silicones, and polyester. By the 1970s, high-tech plastics used in the health and technology fields had been developed. The major raw materials used to manufacture plastics are oil and natural gas. In the plastics production process, crude oil or natural gas is treated in something called a "cracking process." The process converts components of crude oil and gas into hydrocarbon monomers such as vinyl chloride and styrene. Polymers, like ethylene, are created by bonding the monomers into chains creating plastics with many different types of versatile characteristics (Rademacher, 2003).

The non-profit organization, the Ecology Center, has pointed out that besides being prominent in our everyday lives, the production of plastics has been prominent in being the source of hazardous wastes. The Toxic Release Inventory of the Environmental Protection Agency (EPA) has reported that of toxic releases into the air, the plastics industry contributes 14% of the national total in the United States. In addition, seven of the top ten manufacturers releasing toxics into the air do so as a result of the creation of plastic foam products. The chemicals released include acetone, styrene, and benzene, chemicals that have extreme, hazardous properties. The production of plastic resin has its own special issues. Polyethylene terephthalate, commonly referred to as PET, is used in containers for liquids like drinking water (e.g., water bottles). The manufacturing process that goes into the production of a 16-ounce plastic water bottle also produces more than 100 times the toxic emission that emanates from the production of one 16-ounce bottle made of glass (Figure 2.1) (Rademacher, 2003).

Putting aside the chemical threats presented through the waste byproducts of the manufacture of plastics, one must consider the toxic additives from plastics that migrate into food after the

Figure 2.1 This Kansas City photograph illustrates primary air pollution – smoke stack and auto emissions. https://archive.epa.gov/epa/aboutepa/guardian-epas-formative-years-1970-1973.html

plastic products are produced. Direct toxicity exists in the form of exposure to lead, cadmium, and mercury that can be part of plastic materials that are used to encapsulate foodstuffs. These can include antioxidants and heat stabilizers. Chemical interactions facilitated by the additives can produce complex chemical compositions that complicate the dangers presented to humans. Furthermore, wasted plastics sent to landfills present another problem. Because of the stability of plastics over time, they stay in the environment for long periods, particularly if they escaped direct sunlight as a result of being buried in landfills. Since many plastics have antioxidants added to them to resist the effects of acidic contents, the rates of product decomposition are decreased, presenting long-term problems for the environment. In the case of plastics, enhanced technology to provide conveniences, services, and special products to the general public has also led to an ongoing and serious threat to contemporary society that will be slow to subside in the future.

Other Sources of Hazardous Wastes

If we only had to worry about what to do about problems resulting from the manufacture of plastics, control of the threat of hazardous waste would be far more manageable than it really is. The truth is that tens of thousands of chemicals are used every day by industries and businesses in the United States. These industries produce products that are in demand by the general public. They include pharmaceuticals, computers, mobile devices, and automobiles, just to name a few. While not all of these chemicals are toxic, some of these chemicals are released improperly, even though the majority of them are managed well. The previously mentioned Toxic Release Inventory (TRI) of the EPA is an EPA program that tracks the management of toxic chemicals presenting threats to the environment and human health. The TRI National Analysis is developed by EPA on an annual basis for the entire country. The most recent complete database is for the year 2013. For that year, it was reported that production-related hazardous wastes managed by waste-producing entities account for over 25 billion pounds of toxic chemicals (U.S. Environmental Protection Agency, 2013).

The TRI separates information on waste production into seven industry sectors: Chemicals; Primary Metals; Metal Mining; Electric Utilities; Food/Beverages/Tobacco; Paper; and Petroleum. In addition, information is collected on sectors condensed into the category of All Others. For the year 2013, the two top sectors for total waste managed were Chemicals and Primary Metals, accounting for over half of the total hazardous wastes managed in the United States. This proportion has fluctuated slightly over the ten years preceding 2013 but mirrors the proportion reported in 2003. When comparing production-related waste managed between 2003 and 2013, most industry sectors reported a decline in waste managed. The two sectors that reported an increase in the quantity of waste managed since 2003 were Metal Mining and Food/Beverages/Tobacco. Interestingly, between 2012 and 2013, four of the seven sectors demonstrated an increase in the volume of waste managed. Between 2012 and 2013, Metal Mining, as a sector, showed a 22% increase of waste managed, an increase of 332 million pounds of hazardous waste. For that same time period, Chemical manufacturing increased its volume of hazardous wastes produced by 9%, accounting for 856 million pounds of hazardous waste. Tied for third in the increase of the volume of hazardous wastes produced were Electric Utilities (+7%; 109 million pound increase) and Food/Beverage/Tobacco (U.S. Environmental Protection Agency, 2013).

In terms of production-related waste by industry, the chemical industry, by far, accounts for the plurality of hazardous wastes released in the United States. For the year 2013, the chemical industry accounted for 42% of the total hazardous wastes managed. The Primary Metals sector is a distant second in production, representing 11% of the total hazardous wastes produced by industry sectors. Tied for third in production are the Metal Mining sector, the Petroleum sector, and the Electric Utilities sector, representing 7% of the wastes managed for each respective sector (U.S. Environmental Protection Agency, 2013).

The Metal Mining Sector

In the United States, metals that are extracted in U.S. mining operations are used in many products, particularly in the manufacture of industrial equipment and automobiles. Large amounts of hazardous waste are produced in the extraction of minerals and in the process known as beneficiation. This process is critical to removing undesirable minerals (gangue) surrounding metal ore to produce a higher grade of metal. Most of the mining operations reporting the generation of hazardous wastes (88 in total) were concentrated in western states like Nevada, Arizona, and California. Most of these operations involve the mining of copper, silver, and gold. Zinc and lead mining occurs more frequently in Missouri, Tennessee, and Alaska. Generally, metal mining operations result in large volumes of metal and other materials. The unique feature of metal mining is that slight changes in chemical compositions of deposits that are mined can often result in great changes in the volume of toxic chemicals reported across the country (U.S. Environmental Protection Agency, 2013).

One type of waste that is common to mining operations is something called "waste rock." This sometimes is referred to as "overburden material" and is typically removed in efforts to get at the ore deposits in open-pit mines. Waste rock is usually stored at ground level piles at the mine site but can also be stored underwater. A concern regarding wastes from mine operations exists in the form of something called "tailings." This represents the residue left over from the processes used to separate the valuable ore from waste rock. Tailings represent an environmental threat because they often contain toxic metals such as mercury and arsenic. "Mine water" is also an environmental issue at mining sites. It is common for water at mine sites to be closely monitored, and strategies are encouraged to reduce the amount of mine water produced. Gaseous wastes can also be produced by mining activities in the form of particulate dust and sulfur oxides. The types of hazardous wastes produced by mining operations, by and large, depend upon the types of ore being mined and in the processes employed to process the ore (U.S. Environmental Protection Agency, 2013).

The Chemical Industry Sector

The chemical industry sector in the United States represents the highest number of manufacturing entities reporting the generation and management of hazardous wastes. For the year 2013, close to 3,500 chemical manufacturers reported the generation and management of hazardous wastes. These manufacturers produce a variety of products. Besides the production of plastics, they produce synthetic fibers, paints, fertilizers, drugs, and cosmetics. While the chemical manufacturing sector has, over time, consistently been the sector representing the most production-related waste managed, a significant minority (21%) in the sector engaged in source reduction activities in 2013 in an effort to reduce toxic chemicals used in the subsequent generation of hazardous wastes. The chemical manufacturing firms were asked to detail what source reduction methods were used.

The most common category of source reduction activities reported was characterized as "good operating practices." This term represents activities such as the improvement and maintenance of scheduling, recordkeeping, and the alteration of production schedules to reduce the amount of equipment changes. An example of this waste reduction strategy is any effort to increase the batching of products that would lead to the reduction of the frequency of the cleaning of vessels storing chemicals. Another example is processes designed to lessen the chances of spills and leaks (U.S. Environmental Protection Agency, 2013).

Electrical Utilities Sector

Entities responsible for the generation, transmission, and distribution of electrical power make up the electrical utilities sector. A total of 567 electrical generating facilities report to the TRI. Since 2003, net electricity generation (primarily electricity generated using coal and oil fuels) has decreased by 23% since 2003. Due to that factor, production-related waste managed has decreased by 4%. The industry's transition to natural gas has been the reason why its decreases have taken place. Compared to 2003, a much higher percentage of wastes are destroyed through the activation of scrubbers used by electrical utilities, which eliminate acid gases that would otherwise be on-site air releases. Compared to other sectors, a much smaller percentage has initiated source reduction activities to reduce toxic chemical use and waste generation. The modification of equipment and piping has been cited as the main strategy used for source reduction activities in the electrical utility sector (U.S. Environmental Protection Agency, 2013).

Petroleum Industry Sector

The processing of crude oil and natural gas liquids to produce finished petroleum products is the main goal of petroleum refineries in the United States. The key products of the sector are separated into three categories. The first is the category of "fuels," which includes gasoline and kerosene. The second category is a category of "non-fuel" products. Two examples of these products are solvents and asphalt. The final category encompasses petrochemical offshoots like benzene and xylene. An interesting aspect of the sector is that it represents 7% of production-related waste managed, accounting for only 151 facilities. The majority of refineries are located along the Gulf Coast and in Midwestern states near oil fields and ports. From 2003 to 2012, production-related waste in the petroleum industry sector decreased by 17%. Source reduction activities were initiated by 11% of petroleum refineries in 2013. Good operating practices, process modifications, and spill and leak prevention are the primary strategies for the reduction of waste generation in this industry (U.S. Environmental Protection Agency, 2013).

Non-point Source Pollution

A common feature to all of the waste-producing sectors above is that they are all "point source" entities. The U.S. EPA defines point source pollution as any single identifiable source of pollution from which pollutants are discharged (Hill, 1997). Water pollution and unsafe drinking water that restrict activities like fishing and swimming often emanate from unregulated discharges from point sources. The extent to which discharged chemicals are considered harmful depends on the characteristics of the chemicals, the level of concentration of chemicals, and the timing of the release. Besides the various sectors previously described, concentrated animal feeding operations (CAFOs) can be a primary point source of the generation of harmful wastes. Large farms that raise livestock is an example of a concentrated animal feeding operation. The animal waste materials from such operations, if left untreated, form raw sewage that can pollute nearby waterways. The National Pollutant Discharge Elimination System (NPDES) was created under the Clean Water Act to address point source pollution, and requires the procuring of permits from the state and the EPA before effluents are discharged into any body of water. Before discharge, the point source must employ the use of the latest technologies to substantially reduce pollutants before discharge (U.S. Department of Commerce, 2008).

Compared to point source pollution, non-point source pollution is much more difficult to control because pollution comes from many different sources and locations. Runoff from rain or melted snow is a primary source of non-point pollution. In these situations, the runoff is absorbed into the ground or into waterways along with the pollutants. For instance, cars parked in parking lots can contribute to non-point source pollution after a heavy rainstorm by picking up oil left on the asphalt. This runoff will likely run over the edge of the parking lot and empty into a stream that subsequently flows into a larger stream. Ultimately, this polluted water may end up in a river or in the ocean (U.S. Department of Commerce, 2008).

Besides the deleterious effect non-point source pollution has on the environment, it can also have a harmful impact on local industries and local economies. For example, recreational fishing and the fishing industry as a whole can be seriously affected if non-point pollution is the cause of mass die-offs of fish. Continued non-point pollution can result in the decline of property values in areas near waterways. Urban and suburban areas are particularly vulnerable to non-point pollution because of the concentration of asphalt and concrete surfaces in these areas. Water will simply run over the surface and not be absorbed into the soil until it gets picked up by a variety of pollutants along the way. Construction sites are also responsible for non-point pollution when contaminated soil is not carefully handled and contained. Agricultural runoff is also responsible for a significant contribution to non-point pollution, particularly through the utilization of pesticides in the soil around crops.

Discussion

This chapter makes clear that there are many potential sources of hazardous wastes produced in this and other countries. They could range from a "mom and pop" auto body shop, to a farm, to a major manufacturing corporation. The development of plastic represented a key stepping stone toward achieving greater convenience and comfort in modern society. But, at the same time, it also brought us closer to exposure to chemicals in our everyday lives. Much of the toxins released into the air arise from the processes used to create plastics. Additives used in plastics that contain the food and liquids we consume can present an additional threat to health in the 21st century.

The chapter points out that there are under ten industry sectors that produce the vast majority of hazardous substances we can be exposed to. Two such industry sectors, metal mining and chemical manufacturing, have reported increases in the volumes of chemical wastes generated. The residue left over from typical metal mining operations can contain highly toxic metals like arsenic and mercury and must be carefully managed by mine operators, or else they will present harmful results to those exposed to them. Within the chemical industry sector, there are so many categories of chemical waste products that are produced that increasingly greater attention is being paid to attacking the creation of toxic wastes through source reduction programs. Greater efforts like this are needed in the electrical utilities sector. To help ensure a safe environment, effective monitoring of the petroleum industry sector remains critical, particularly in geographic regions like the Midwest and the Gulf states where refineries are concentrated. A final concern underscored in the chapter is that presented by waste streams that have no single industry source: non-point pollution. This type of pollution is the most difficult to control because there are so many possible sources. A major threat here is from liquid "runoff," especially from agricultural operations.

Discussion Questions

1) The development of synthetic materials that help to make our lives more comfortable and enjoyable has also been the source of some resultant environmental concerns. Do you feel these concerns are so threatening that we need, as a society, to assess whether the value of these processed products outweighs the risks? Why or why not?
2) One way of addressing the issue of lessening the impact of hazardous substances in our lives is to pay more attention to the beginning of the manufacturing process rather than to the endpoint. One of these methods is "source reduction." How viable do you think this approach is in helping to minimize the risk associated with waste exposure? Why do you think some industries are hesitant to embrace this approach? What methods can be employed to encourage industries to increase source reduction activities?

References

Hill, M.S. (1997). *Understanding environmental pollution*. Cambridge, UK: Cambridge University Press.
Rademacher, D. (2003, November 15). *Plastic, not so fantastic*. The Ecology Center. Retrieved from http://ecologycenter.org/terrainmagazine/winter-2003/plastic-not-so-fantastic/
U.S. Department of Environmental Protection. (2013). *Toxic release inventory program*. Retrieved from http://www.epa.gov/toxics-release-inventory-tri-program
United States Department of Commerce. (2008). *Nonpoint pollution*. National Oceanic and Atmospheric Administration. Retrieved from http://oceanservice.noaa.gov/education/kits/pollution/04nonpointsource.html

Chapter 3

The Legitimate Treatment of Hazardous Waste in the United States
What Was Done in the Past and What Is Done Today

While this book is dedicated to examining the illegal disposal of substances that harm the environment and the enforcement of laws to control this, it is important to note and understand the manner in which these materials are handled legally. In most cases, fortunately, these substances are handled properly in the United States. However, certain pressures and circumstances have made it more likely that a growing percentage of cases will involve the improper and illegal transportation, storage, and disposal of hazardous wastes. This phenomenon can be partly explained by the expansion of production by corporations plus the desire of these corporations to sport new markets. The result has been an increase in the volume of hazardous wastes since the end of World War II. The corporate world has its ever-present pressure for the cultivation of new markets leading to the creation of new waste materials and the growing acceptance of embracing improper and illegal methods of disposal. The petroleum and chemical industries have rationalized that because of the demand for their products, society, in general, should take greater responsibility for the legitimate disposal of hazardous wastes (Brownstein, 1981).

Complicating this matter is the fact that competition has become fierce in these industries, raising the stakes for optimizing efficiency in production activities. It stands to reason that the industries feeling the brunt of pressure are those that generate the greatest volumes of hazardous wastes. So, in essence, these industries are greatly impacted by financial pressures. And this does not account for the special pressures felt by small businesses. In these firms, owners and managers are forced to dig into their limited assets in an effort to operate according to the standards of disposal regulation. Compounding this problem is the fact that since the end of World War II financial expenses associated with the legitimate disposal of hazardous waste has risen exponentially. This then has become the primary driving force for the illegal disposal of hazardous waste as an answer.

Historically, generators of hazardous waste have had a number of options for the legal disposal of these wastes. At one time, disposing of hazardous waste in landfills was permitted and was the leading method of hazardous waste disposal in the United States. That time is now gone unless landfills are properly equipped, by legal standards, to accept such wastes. Replacing the landfilling of hazardous wastes was a new method, which was the incineration of hazardous wastes at high temperatures. Adding to that method, we now have the means of treating hazardous wastes biologically or chemically to destroy the noxious elements of the substances contained within these wastes. In some cases, these wastes can be treated in a way that converts them to water or carbon dioxide. Adding to this, treated properly, these wastes can be recycled and reused. While controversial, deep well injection can become the resting place of hazardous wastes if those wells are considered deep enough by law (Rebovich, 2015; Westat Inc., 1984; Reimers, 1985; Sarokin, Muir, Miller and Sperber, 1985).

With regard to the legitimate treatment of hazardous wastes, the greatest stumbling block is the burgeoning cost inherent in the proper treatment of hazardous wastes. For one 55-gallon drum of material containing hazardous waste, waste generators must pay anywhere between $100 to close to $1,000 depending on the chemical makeup of the substance. However, before a waste generator or treatment facility treats a hazardous waste, the waste must be stored. This chapter's content begins with legitimate waste storage and walks the reader through the various established means of legitimately treating and disposing of hazardous wastes, including emerging technologies to treat and dispose of wastes more effectively and efficiently.

Storage

Whenever a hazardous waste has been transported to a treatment/storage/disposal (TSD) facility, it must be placed in temporary storage. Such a facility can use several methods for this type of storage. One method is to simply put the wastes into sealed containers. A container of this kind that contains any type of hazardous waste is both sealable and portable. The most common type of container for such storage is a 55-gallon drum. Depending on the type of waste, the drum can be either metal or plastic. By no means can corrosive wastes be stored in these types of containers for that would lead to a deterioration of the container walls. A second type of waste containment method is to store the wastes in a containment building. It is important that such containment buildings do not come in contact with any other building and must be completely enclosed. In essence, a containment building is a freestanding building with the floor, four walls, and a roof (Layton, 2006).

Besides these two types of containment methods, there are several other options. One such option is a surface impoundment. This is an in-ground structure that is either a natural or a man-made depression in the ground. It is important to note that certain requirements must be met for this method to be legally implemented. Principally, the hazardous wastes in a surface impoundment must be prevented from leaking into the ground. This is usually accomplished by lining the surface impoundment with heavy plastic. Storing hazardous waste in tanks is another option. The material the tank is made of is critical. Such tanks are usually made of steel, fiberglass, concrete, or, sometimes, plastic. It is up to the tank custodian to determine if the tanks are completely enclosed or are open topped. Only hazardous wastes that do not give off gases can be stored in open-topped tanks. A final option is represented through the use of "waste piles." These are simply mounds of hazardous waste piled at ground level. While the wastes are completely open, it is mandatory for this type of storage to have no contact with the ground. Therefore, in the waste pile method, there must be an impenetrable material that separates the waste from the ground. Again, any wastes that emit gases cannot be stored through the use of waste piles (Layton, 2006).

An important consideration in the proper storage of hazardous wastes is that "incompatible wastes" not be stored next to each other. If stored next to each other, there are certain chemical wastes that will lead to a chemical reaction that can cause a fire or an explosion. Due to the volume of hazardous wastes that are stored at some facilities, it can become very difficult to ensure that incompatible wastes are not always stored away from each other. Rebovich notes, in his study of hazardous waste crime (2015), that the ordering by unethical facility supervisors to yard workers to intentionally position incompatible wastes in close proximity often proves to be the first step in a criminal conversion process leading to more direct illegal disposal of wastes by the workers. Those storing hazardous wastes are required to clearly mark each storage structure with the types of substances contained in the structure. This is done in an attempt to prevent any mistakes that might be made on the types of wastes contained and, consequently, avoid accidents that may be catastrophic.

Combustion

Once hazardous wastes have been temporally stored, they are then ready to undergo some type of treatment process (with the exception of proper landfilling) and then final disposal. The most widely used type of combustion method is combustion through the use of an incinerator, followed by the use of boilers and industrial furnaces. Boiler and industrial furnaces (BIFs) are primarily used to burn hazardous waste for the purpose of energy and material recovery potential. Through this process, waste treatment is actually seen as a secondary benefit. Industrial furnaces burn waste for both material recovery and energy, while boilers primarily burn the waste for the purpose of energy recovery. A boiler is considered any enclosed device that employs the use of controlled flame combustion for the purpose of recovery and exportation of energy. This usually takes the form of heated fuel, heated gases, or simple steam. An industrial furnace uses thermal treatment to recover materials or energy. There is a wide array of different types of units that can be used as industrial furnaces. They include blast furnaces, coke ovens, and cement kilns. More complex forms include titanium dioxide chloride process oxidation reactors and halogen acid furnaces (U.S. Environmental Protection Agency, 2015).

While the use of incinerators can produce some energy or material recovery, the primary purpose of the use of incinerators is to burn hazardous waste for destruction and treatment purposes. Effective incineration of hazardous waste not only destroys the toxic organic components within it; it also is responsible for significantly reducing the original volume of the waste. Any hazardous waste that contains elements of metal is not suited for incineration since those types of waste will not combust through the incineration process. Like BIFs, there are many different types of hazardous waste incinerators. They include rotary kilns, fixed hearth units, and liquid injection units (U.S. Environmental Protection Agency, 2015).

One of the more popular methods for incineration is the use of rotary kilns. These kilns must be able to withstand very high temperatures to incinerate hazardous wastes. For this method of incineration to be done effectively, the temperature should reach or go above 1,800°F. Such temperatures in the kiln are often maintained by injecting natural gas or other supplemental fuels into the chamber. The process requires that liquid wastes be pumped into the kiln through nozzles. The liquids are then reduced into droplets of one-millionth of a gram. If the wastes to be destroyed are in solid form, they are introduced into the kiln either in containers or in bulk form employing the use of conveyors or gravity feed systems. The solid wastes are dispersed as the kiln rotates to ensure that the wastes are totally exposed to the high temperatures. In effect, it is the same concept employed in a much simpler form in the use of an average dryer used to dry clothes. As the kiln is rotated, oxygen is injected into the film to increase combustion through the use of large fans. As the wastes are converted into gases through this process, they pass into an afterburner. Supplemental fuels are then injected into the afterburner in which temperatures are raised even higher, sometimes over 2,000°F. Gaseous and atomized organic compounds are, thus, transformed into atoms that reconnect with oxygen from the air. After this process is complete, the end product can be non-hazardous chemicals such as carbon dioxide (Environmental Technology Council, 2015).

Each incineration system must also have an air pollution control system (APCS) within which generated gases are cooled and cleaned. Any small solid matter (particulates) is removed in this process. These incinerators also have something called "waste feed cut offs" or (WFCOs). These cutoffs are imperative to the integrity of the entire incineration operation, for they are responsible for halting the introduction of waste into the incinerator if certain stringent parameters are not met during the process (e.g., temperatures dropping below mandatory levels). Any inorganic ash that results from the process is meticulously examined to ensure that no hazardous contaminants exist. Chemical stabilizers are then mixed with the ash. Once it is ascertained that this ash has met EPA standards, it is placed in a landfill that has met EPA minimum technology requirements. Such landfills are equipped with two liners and a leachate collection system. This then becomes the endpoint of the hazardous waste incineration process, a methodical stepwise process that has ensured that any end product is rendered innocuous. Based upon the description of this process, it becomes clear why it is an expensive one (Environmental Technology Council, 2015).

The EPA has determined that destroying hazardous wastes in incinerators is the best demonstrated available technology (BDAT) for most organic waste. When properly administered, the incineration of hazardous waste not only destroys the toxic elements within it but also significantly reduces the waste volume. On average, the incineration process is responsible for reducing the solid mass of hazardous waste by up to 85% and the volume by up to 96%. The incineration does not eliminate the need for landfilling, but markedly reduces the total amount that winds up in landfills as the final destination. The process is particularly instrumental in the effective destruction of clinical wastes and wastes that contain pathogens that must be destroyed at very high temperatures (Cooper, 2015).

Secure Landfill Disposal

As mentioned earlier, another popular method of handling hazardous wastes is to dispose of them in a landfill. There is an important caveat, though. They have to be special landfills that meet strict regulations. These federal and state standards are in place for landfills for the protection of public health and the environment. The stated regulations are part of the Resource Conservation and Recovery Act (RCRA). These regulations are directed at the proper design, location, construction, and operation of a landfill. The regulations also address the final closure of a landfill. All secure landfills must have RCRA permits that are directed at ensuring that the preceding requirements are met. Many individual state requirements are even more stringent than the federal requirements. For instance, some states require groundwater monitoring wells. Many states have restrictions on the disposal of radioactive wastes at secure landfills. In addition, operators of secure landfills can expect on-site state inspector visits periodically. It is important to note that state requirements must be as stringent, or more stringent than federal regulations. They may never be less protective (Environmental Technology Council, 2015).

The primary requirement of hazardous wastes landfills is that they have double composite liners containing the wastes. In addition, they must have a leachate collection system both above and between the liners. The landfills are also required to have a leak detection system. It is mandatory that these leak detection systems are able to detect any leakage between the liners at the earliest possible time. The detection systems must also be capable of collecting and removing any leakage when it occurs. If leakage does occur, it must be removed and effectively treated to ensure that the groundwater is protected. These types of landfills are also required to control "run on" and "run off." The diversion of runon is implemented to ensure that there is little or no erosion to the landfill. Runoff is a different problem and addresses the issue of precipitation. To prevent migration off site, runoff precipitation must be collected and managed properly. It is mandatory for landfill operators to determine if the runoff contains hazardous waste and, if so, properly treat it. Wind dispersal is another potential problem requiring landfills to be covered. Hazardous wastes that are shipped to secure landfills are under the purview of the federal manifest system. This is in place to make sure that regulators are able to trace the waste from "cradle to grave." The testing of wastes that are shipped to such landfills must be done to ascertain what the elements are of substances. This is done to ensure that incompatible wastes are not mixed together in the landfill, reducing the chances of explosions or fires (Environmental Technology Council, 2015).

A variety of technologies are used by companies to ensure that the landfilled wastes meet EPA treatment standards. Once again, these standards are based on the performance of the Best Demonstrated Available Technologies (BDAT). The two most common methods for doing this are stabilization and neutralization. These processes are used on both hazardous and non-hazardous sludges, liquids, soils, and slurries. Materials derived from living organisms or hydrocarbons are treated differently. These organic wastes are chemically oxidized before they are disposed of and landfilled. Metal-bearing wastes undergo stabilization techniques to convert them into an insoluble solid material (Environmental Technology Council, 2015).

The EPA is responsible for establishing minimum standards for landfill security and inspections. In addition, federal standards have been put in place directed at how landfill personnel should be trained. The actual sighting of secure landfills also must meet restrictions to avoid unstable conditions in the area that may result in problems connected with conditions like flooding. Closing down a secure landfill is also strictly regulated. Once a secure landfill is closed, the owners of the landfill must perform federally mandated monitoring and maintenance activities for a minimum of 30 days after closure. Prior to closure, secure landfill owners must meet certain financial responsibility requirements so that adequate funds are available to effectively implement monitoring and maintenance activities during the 30-day period after landfill closure.

To reduce the volume of hazardous waste that is sent to landfills, many firms use source reduction and recycling methods. Further, firms will send industrial wastes to secure landfills that are not considered "hazardous" by RCRA standards, yet contain low levels of toxicity (Environmental Technology Council, 2015).

New Technologies to Treat Hazardous Waste

Since the beginnings of the use of combustion and secure landfilling for the disposal of hazardous wastes, research has been continuing to attempt to find new and more efficient ways of dealing with the problem of hazardous waste. Some newer technologies used include UV/oxidant processes which employ the use of peroxide or ozone for the treatment of the wastes. Another method is the molten-salt combustion technique. In this process, a pool of molten salts ignites the combustion of hazardous substances below normal ignition levels.

One of the newer methods with the most potential for the future has to do with bacteria that can eat hazardous waste. Early studies of this approach were conducted by Dr. John Coates and Dr. Laurie Achenbach of Southern Illinois University. These two scientists discovered over 40 different bacteria that can break down certain toxic wastes into a component of ordinary salt and other harmless chemicals. They were able to find these bacteria in a variety of locations, including sediments from freshwater sources and ocean sources, as well as soil samples collected from Antarctica. The microbes were found to be able to convert iron dissolved in wastewater into a solid form through the chemical reaction of oxidation. This process involves extracting an electron from each atom of ferrous iron, preventing it from dissolving in water, and transforming it into a solid form. The electrons are used by the microbes as energy to create cellular activities. Like a magnet, the newly formed ferric iron connects to toxic compound atoms. In the final step of the process, solid iron is collected out of the soil or water and harmful substances removed. To elevate the treatment process, specific carbon sources must be added. The researchers were able to elevate the waste-eating qualities of the bacteria by adding inexpensive carbon sources to the soil or water meeting cleanup (American Society of Microbiology, 2006).

More recent technology has been developed focusing on nuclear waste in underground repositories. Scientists at the University of Manchester have addressed the potential problem of buried nuclear wastes being subject to change over time that might threaten the general ecosystem (Bassil, Bryan and Lloyd, 2014). Radioactive waste that is encased in concrete prior to disposal in underground vaults is susceptible to reactions if groundwater reaches the waste materials and reacts with the cement to become highly alkaline. A chemical reaction can result that triggers the breakdown of cellulose-based materials present in the wastes. One product of this activity is something called isosaccharinic acid or ISA. This can bind to elements of the wastes (i.e., radionuclides), making them more soluble with the potential of flowing out of the vaults presenting a direct threat to surface environments. The scientists' research uncovered bacteria that thrive in such alkaline conditions (i.e., extremophile bacteria). This bacteria was found to use ISA as a source of food and energy, thus significantly reducing the threat of drinking water and food chain contamination from the buried nuclear wastes (Bassil, Bryan and Lloyd, 2014; Brewin, 2014; Scutti, 2014).

Discussion

Chapter 3 explains the types of methods that are used to treat and dispose of hazardous wastes and how those methods have evolved over time. In the past, such wastes were legitimately disposed of in landfills, regardless of the structural characteristics of those landfills. As society became more aware of the environmental dangers posed by unfettered dumping of chemical substances into landfills, restrictions were put into place requiring that special standards be met. Greater restrictions were also placed on the manner in which hazardous wastes could be stored to prevent leakage and combustible chemical reactions.

Presently, the method used most frequently by those desiring to legally dispose of hazardous waste is combustion. In some cases, special furnaces are used for this method. But, the most popular legitimate method to destroy such wastes, through combustion, is the use of incinerators that burn the wastes at very high temperatures. The use of a rotary kiln is the most common unit used for incineration. Once again, however, the proper use of incinerators within legal standards can be a complex operation. Scientists are constantly researching new alternatives to presently used methods in the interest of making the employment of treatment/disposal methods as safe as possible for the environment as well as for those conducting the procedures. Recent breakthroughs have been seen in the area of biologically produced treatment/disposal methods that include the exploration of types of bacteria that can actually consume chemical wastes. Research in this field is ongoing as we continue to search for better ways to treat and dispose of dangerous chemical wastes.

Discussion Questions

1) While the utilization of incineration facilities to dispose of hazardous waste has been seen as successful, a problem remains. Where do we put them? What kind of difficulties do you believe are associated with the siting of such facilities? Would you want such a facility built near where you live? Why or why not? Discuss why there might be a societal/economic unfairness in where such facilities are eventually sited.
2) Major advances have been made in the biological treatment of hazardous waste. How do you feel about this? Do you see any dangers in helping to create new forms of "super" bacteria that have the power to "eat" chemical wastes? What do you foresee as other fields in which technological research can help improve our quest to effectively treat/dispose of hazardous waste?

Case Study: The Use of Bioremediation to Neutralize Hazardous Waste

In Buncombe County, North Carolina, there are at least 30 hazardous waste sites as a result of manufacturing operations that had closed down and abandoned buildings and properties that contained toxic materials generated by the firms. Writer Max Hunt has reported that many of those firms had been operating in the years before the U.S. government took a harder stand against environmental pollution by enacting strict regulatory and criminal laws, and enforcing them, in an effort to protect our environment. The contamination left behind by these companies became the headache of local government and residents seeking to clean up what was left behind, a mammoth task that is both time-consuming and expensive. One example is in Swannanoa, NC, where the Chemtronics waste site has remained on the EPA's National Priorities List since the year 1982. Another is located in South Ashville, NC, where extraordinary delays in the cleanup of the CTS Superfund site led to outrage on the part of community activist groups (Hunt, 2017).

Sometimes the dire need to solve a protracted problem can lead to the development of innovative new methods to solve that problem. Such was the case in North Carolina, where innovative strategies collectively known as "in situ remediation" have been seen as the answer to cleaning up hazardous waste sites more effectively and efficiently. This method is quite different from physically removing hazardous wastes from contaminated properties. Instead of removal, in situ remediation takes advantage of certain properties of chemicals and bacteria to transform toxic substances into innocuous ones. In situ remediation is a form of bioremediation. According to the EPA, bioremediation is an engineered technology that modifies environmental conditions (physical, chemical, biochemical, or microbiological) to encourage microorganisms to destroy or detoxify organic and inorganic contaminants in the environment. The process can be applied above ground in farms, tanks, biopiles, or other treatment systems (referred to as ex situ) or below ground in the soil or groundwater, referred to as in situ. This latter strategy was employed in the aforementioned two waste sites (Hunt, 2017).

The approach at Chemtronics relied upon this form of bioremediation in which nutrients and oxygen were injected into waste substances. The approach at the CTS site was a bit different. There, a mixture of chemical oxidizers was injected into soil as a way to dissolve the molecular structure of the contaminants. At a third North Carolina waste site, grass infused with special "pollution eating" bacteria were planted on the grounds of the site. Earlier efforts used at the Chemtronics site to clean up the waste there relied on traditional "pump and treat" activities to remove the waste site. Authorities came to realize such strategies did not fully address contamination in surrounding soils and bedrock structures and only effectively dealt with less than a quarter of the contamination. In addition, there were added expenses for electricity needed to run extraction wells, treatment processes, the discharge of untreated groundwater into sewer systems, and the maintenance of discharge lines (Hunt, 2017).

Officials went through several years of on-site pilot studies at Chemtronics before deciding on in situ remediation. A consulting firm was brought in to design a matrix of injection wells on the property close to contaminated groundwater plumes and then injected lactate solutions and emulsified vegetable oil along with pollution-eating bacteria. This represented the start of the endeavor to attack the large concentrations of fuel oil mixed with trichloroethylene (TCE), a known human carcinogen, on the Chemtronics property. After the first approach was implemented, other approaches followed. One involved using electrical resistance heating underneath where the original factory was located. This heating process was used to boil the groundwater that the TCE was converted into a vapor. Another part of the grounds where the factories stood was injected with a chemical that created a change in the electron states of various compounds. In essence, the process converted TCE molecules into innocuous substances. The two-pronged approach was designed to attack TCE concentrations with heat closer to the surface and chemical treatment for contaminants lower beneath the ground (Hunt, 2017).

One difficulty in this bioremediation method was actually making sure solutions injected into the ground had direct contact with the contaminated groundwater plumes. The hurdles lay in achieving adequate distribution of the oxidants that was highly dependent on the availability of precise site-specific characteristic data. Electrical resistance heating also had its challenges, the primary one being ensuring the effectiveness of monitoring vapor emissions to prevent unintended harm to residents living near the site. During the early stages of the processes, ambient outdoor air testing found elevated concentrations of TCE readings with subsequent later readings registering lower concentrations. Additionally, authorities had to be on the alert for other unintentional harmful byproducts of the in situ remediation procedures. It was found that some compounds, during the process, could possibly be broken down into more toxic byproducts, like the conversion of TCE to vinyl chloride. Consequently, those that operate these processes had to have a clear appreciation of how the soil could react to injected elements. For the Chemtronics situation, it was found that the use of bacteria could have the side effect of acidifying the soil and creating a leachate of potentially dangerous elements like magnesium from rock. The use of a neutralizing agent, in this case, sodium carbonate, was injected to build a buffer allowing the bacteria to multiply (Hunt, 2017).

Initial signs were that the use of in situ remediation in the Chemtronics case had been successful with EPA projections that TCE contamination over the years will be reduced by over 90%. It has been noted that this type of bioremediation has also been effectively employed at former weapons facilities to eradicate ground contamination. While there has been some concern expressed regarding degraded byproducts of pollutants, long-term study results at other sites have demonstrated that in situ remediation is, in general, a safe method to address the problem of groundwater contamination and should increase in popularity in the future (Hunt, 2017). However, circumstances in other similar bioremediation cases can give one pause.

In February of 2017, on the other side of the world, a collision between two ships off the coast of India resulted in leakage of heavy furnace oil, otherwise known as bunker oil, into the sea, contaminating it with as many as nine heavy metals. These metals included chromium, cadmium, and selenium, along with polyaromatic hydrocarbons and asphaltenes. Indian Institute of Technology experts initially claimed that bioremediation could remove the lighter aromatic hydrocarbons, but the heavier hydrocarbons and heavy metals that are non-biodegradable would remain in the environment for decades (Chaitanya, 2017).

Flaws were detected in how the large-scale bioremediation process was conducted in this case. The oil sludge was mixed with clean sand and bioremediated through a process that adds inoculum and nutrients. This process resulted in an actual increase in the volume of waste, which ended up violating India's three R concept (Reduce, Reuse, and Recycle) for effective waste management in that country. In addition, the bioremediation processes were to be conducted following the guidelines of India's Central Pollution Control Board for oil-contaminated sites. These guidelines were initially created for in situ remediation. In this particular case, oil was transported from beaches, and, according to law, bioremediation should have been conducted ex situ in concrete tanks. The proper procedures here would have been incineration in a cement plant or incineration in a controlled manner in a hazardous waste incinerator. Since the oil had been mixed with clean sand in a port area, some precautions should have been taken, like keeping the shallow groundwater table safe from contamination, preventing rainwater runoff, and close monitoring of the bioremediation process. This monitoring should have been conducted by a third party to ensure that the oil was degraded to the greatest extent without transferring the pollution to soil or in areas in which substances could eventually flow back into the ocean (Chaitanya, 2017).

In summation then, it can be said that bioremediation methods used to address the dangers presented by hazardous wastes that are either abandoned or otherwise disposed into the environment can be the answer for the provision of sound hazardous waste treatment methods for the present and future. However, it is important to carefully consider the possible unintended effects of failing to follow proper protocol and guidelines in the implementation and administration of what can be effective but complex alternatives for hazardous waste treatment.

References

American Society of Microbiology. (2006). *Presto! Bacteria turn toxic wastes into salt component and oxygen*. Retrieved from http://archives.microbeworld.org/news/articles/microbial_magic.aspx

Bassil, N., Bryan, N., & Lloyd, J. (2014, July). Microbial degradation of isosaccharinic acid at high pH. *The ISME Journal, 9*(3), 310–320.

Brewin, K. (2014, September). Scientists discover hazardous waste-eating bacteria. University of Manchester. Retrieved from http://www.manchester.ac.uk/discover/news/scientists-discover-hazardous-waste-eating-bacteria

Brownstein, R. (1981). The toxic tragedy. In R. Hader, R. Brownstein & J. Richard (Eds.), *Who's poisoning America: Corporate polluters and their victims in the chemical age?* San Francisco, CA: Sierra Club Books.

Chaitanya, K. (2017, February 21). Sludge DNA reveals heavy metals. *The New Indian Express*. Retrieved from http://www.newindianexpress.com/cities/chennai/2017/feb/21/sludge-dna-reveals-heavy-metals-1572903.html

Cooper, S. (2015, April 20). Incinerating hazardous waste. *Environmental Protection*. Retrieved from https://eponline.com/articles/2015/04/20/incinerating-hazardous-waste.aspx?admgarea=ht.waste

Environmental Technology Council. (2015). *High temperature incineration*. Retrieved from http://www.etc.org/advanced-technologies/high-temperature-incineration.aspx

Environmental Technology Council. (2015). *Secure landfill disposal*. Retrieved from http://www.etc.org/advanced-technologies/secure-landfill-disposal.aspx

Hunt, M. (2017, February 23). In situ remediation could revitalize hazardous waste sites. *Mountain Express*. Retrieved from http://mountainx.com/news/in-situ-remediation-could-revitalize-hazardous-waste-sites/

Layton, J. (2006, October 6). How do facilities store hazardous waste? *Science*. Retrieved from http://science.howstuffworks.com/environmental/energy/storing-hazardous-waste.htm

Rebovich, D. (2015). *Dangerous ground: The world of hazardous waste crime*. New Brunswick, NJ: Transaction Publishers.

Reimers, R. (1985). Emerging technologies of hazardous waste. In *Source reduction of hazardous waste*. Trenton, NJ: N.J. Department of Environmental Protection.

Sarokin, D.J., Muir, W.R., Miller, G.G., & Sperber, S.R. (1985). *Cutting chemical wastes: What 29 organic chemical plants are doing to reduce hazardous wastes*. New York: Inform Inc.

Scutti, S. (2014, September 10). Hazardous waste-eating bacteria are evolving over time, may help prevent leaks from underground dumps. *Medical Daily*. http://www.medicaldaily.com/hazardous-waste-eating-bacteria-are-evolving-over-time-may-help-prevent-leaks-underground-302264

U.S. Environmental Protection Agency. (2015, November). *Wastes – Hazardous wastes-combustion*. Washington, DC: U.S. Environmental Protection Agency. http://www3.epa.gov/epawaste/hazard/tsd/td/combustion.html

Westat Inc. (1984). *National survey of hazardous waste generators and treatment, storage and disposal facilities regulated under RCRA in 1981*. Washington, DC: Environmental Protection Agency, Office of Solid Waste.

Chapter 4

The Legal Environment of Environmental Crime

Introduction

Legal protection of the environment is effected through a coordinated scheme of federal and state statutes and regulations that seek to prevent or minimize the hazards of numerous types of pollution. The imposition of criminal penalties is a powerful means of protecting the environment because the mere presence of those promotes compliance with environmental regulations, aids enforcement efforts, and deters future crime.

This chapter includes a description of the major federal laws and regulations designed to protect the environment, selected state laws that provide additional protection of regional and local environmental concerns, the types of criminal conduct (actus reus) prohibited by those laws and regulations, the mental state (mens rea) required of the offender, and the principles governing the imposition of liability upon offenders.

Federal Laws and Regulations

Figure 4.1 The U.S. Supreme Court building. www.supremecourt.gov/about/courtbuilding.aspx

Federal legal protection of the environment (Figure 4.1) can be categorized into seven areas – air, water, solid waste, hazardous waste, chemicals and toxics, pesticides, and wildlife protection. The federal laws in each area include provisions for the imposition of criminal penalties upon offenders and, as well, contain additional civil and administrative remedies for a violation of the applicable statute or regulation.

Air

The Clean Air Act (CAA) was enacted in 1963 for the stated purpose "to protect and enhance the quality of the Nation's air resources" (42 U.S.C. §7401[b]). The Act authorizes the Environmental Protection Agency (EPA) to establish National Ambient Air Quality Standards (NAAQS) and directs the states to establish a State Implementation Plan (SIP) for the reduction of air pollution. The Act further directs the EPA to establish National Emission Standards for Hazardous Air Pollutants (NESHAP). Essentially, the EPA establishes the national air pollution control standards that states are responsible for enforcing at state and local levels through EPA-approved SIPs.

Individuals and entities can violate the CAA in a number of ways: (1) by emitting "criteria pollutants" (carbon monoxide, sulfur oxides, nitrogen dioxide, particulate matter, ozone, and lead) in violation of standards set forth in state plans; (2) by emitting hazardous air pollutants in violation of NESHAP; (3) by emitting pollutants in violation of the New Source Performance Standards (NSPS), which apply to major sources of air pollution; (4) by emitting sulfur dioxide in excess of limiting standards; and (5) by emitting or producing an unauthorized ozone-affecting substance.

The criminal violations of the CAA are set forth in 42 U.S.C. §7413(c). Essentially, the Act imposes criminal sanctions upon those who engage in the conduct set forth in the preceding paragraph; interfere with or fail to install EPA monitoring devices; fail to notify or file reports as required by the CAA; make false statements or certifications or who omit material information from any report, plan, or other documents; and those who fail to pay fees that are owed to the United States.

Water

The Clean Water Act (CWA) was enacted in 1977 with the objective to "restore and maintain the chemical, physical, and biological integrity of the Nation's waters" (33 U.S.C. §1251[a]). The Act is designed to accomplish this purpose by prohibiting the discharge of any pollutant into "navigable waters" of the United States without a permit issued under the National Pollutant Discharge Elimination System (NPDES) permit program.

The meaning of "navigable waters" has been the subject of considerable confusion. Initially, the term was interpreted broadly. In 2006, however, the U.S. Supreme Court stated two tests that were more restrictive (though not less confusing): (1) whether the waters had a "significant nexus to waters that are navigable in fact or that reasonably could be so made" or (2) waters that could be excluded because they are either not "relatively permanent" or wetlands that have no "continuous surface connection" to waters of the United States (*Rapanos v the United States*, 547 U.S. 715, 717, 733).

Although the CWA and EPA regulations apply to the discharge of pollution from point and non-point sources, criminal sanctions are imposed only for discharges from point sources, which are defined by the Act to include

> any discernible, confined and discrete conveyance, including but not limited to any pipe, ditch, channel, tunnel, conduit, well, discrete fissure, container, rolling stock, concentrated animal feeding operation, or vessel or other floating craft, from which pollutants are or may be discharged. (33 U.S.C. §1362(14))

The CWA also prohibits the discharge of oil and hazardous substances in harmful quantities into U.S. waters, dredging and fill activities into navigable waters without a permit from the Army Corps of Engineers, and requires owners and operators of point sources to monitor and submit reports of pollutant discharges. EPA officials can additionally conduct regulatory searches to inspect the books and records of monitoring and discharge activities by point sources.

Criminal penalties are imposed for violations of the various statutory and permit requirements, as well as the falsification or misrepresentation of material information on required reports and the tampering of monitoring equipment.

Rivers and Harbors Act of 1899

Also known as the Refuse Act, the Rivers and Harbors Act prohibits the introduction or discharge into any navigable waters of the United States of any refuse of any kind without the appropriate EPA permit. The Act also extends to the discharge of refuse into any tributary or to the banks of a tributary where the refuse will eventually be introduced into the navigable water. "Navigable water" has the same meaning under this Act as under the CWA.

Refuse has been defined to include "all foreign substances and pollution" but it does not include liquid that flows from streets and sewers (*United States v Standard Oil Co.*, 384 U.S. 224, 230 [1966]).

Safe Drinking Water Act

The Safe Drinking Water Act was enacted in 1974. It addresses concerns with the safety of public drinking water. The Act authorizes the EPA to establish regulations for the underground injection of contaminants in public water systems, the levels of lead in water coolers, and the tampering with public water systems. Tampering includes the introduction of a contaminant into a public water system or interfering with the operation of the system with the intent to cause harm.

Solid Waste and Hazardous Waste

Resource Conservation and Recovery Act (RCRA)

The RCRA was enacted in 1976 to provide for the regulation of solid waste and hazardous waste. More specifically, the Act regulates the generation, treatment, storage, transportation, and disposal of solid or hazardous waste. Solid waste includes garbage, refuse, sludge from treatment plants, and discarded materials. Hazardous waste is a type of solid waste that "because of its quantity, concentration, or physical, chemical, or infectious characteristics" may cause death or serious illness and poses a substantial hazard to human health if not properly managed (42 U.S.C. §6903).

The RCRA imposes criminal penalties for the treatment, storage, or disposal of hazardous waste or used oil not identified as hazardous waste without a permit; the transportation of hazardous waste to a facility that lacks a permit; violation of the terms or requirements of a hazardous waste permit; making false representations in, or omitting material information from, applications, manifests, permits, or other documents filed, maintained, or used to comply with regulations governing hazardous waste; transporting without a manifest any hazardous waste or used oil not listed or identified as such waste or oil; and exporting hazardous waste without the consent of the receiving country or in violation of international agreements (42 U.S.C. §6928[d]). Criminal violations are felonies, and the penalties are substantial.

Civil penalties are also imposed for violations of the Act that involve both hazardous waste and non-hazardous solid waste.

Comprehensive Environmental Response, Compensation, and Liability Act (CERCLA)

Four years after the enactment of the RCRA, Congress enacted CERCLA. Unlike the RCRA, which regulates the transportation, storage, and dumping of hazardous solid wastes, the purpose of CERCLA is to control the environmental problems presented by abandoned or inactive hazardous waste disposal sites. It does not, therefore, deal with ongoing and active hazardous waste sites; the environmental issues raised by active sites are regulated by the RCRA and other environmental laws.

Although CERCLA is remedial in nature and the thrust of the Act is the imposition of civil damages for violations of the Act (*see*, 42 U.S.C. §9607), CERCLA does impose criminal sanctions for the failure to report immediately the release of a hazardous substance, providing false or misleading information, failure to notify the EPA of the existence of an unpermitted site dealing with hazardous substances, the destruction or falsification of records, and the submission of false claims for Superfund reimbursement (42 U.S.C. §§9603, 9604, 9612).

Toxic Substances

Toxic Substances Control Act (TSCA)

The Toxic Substances Control Act was enacted in 1976 to regulate chemical substances that pose a risk to health or the environment and to authorize the EPA to take enforcement action in the event of imminent chemical hazards.

Like most environmental laws, the Act imposes both civil penalties and criminal sanctions. Under the Act, it is a crime to refuse to comply with provisions of the Act that require testing of chemical substances; to manufacture, process, or distribute chemical substances or mixtures without required testing or in violation of the EPA's restrictions concerning the concentrations or distribution levels of the chemical; to fail to comply with orders issued concerning imminent chemical hazards; use for commercial purposes any chemical substance manufactured, processed, or distributed in violation of the Act or EPA regulations; to fail or refuse to maintain and submit required documents, reports, and records or to refuse to permit copying of records; to fail or refuse to permit entry for inspection purposes; and the violation of laws and regulations pertaining to lead-based paint and lead contamination (15 U.S.C. §2614, 2689).

Federal Insecticide, Fungicide, and Rodenticide Act (FIFRA)

FIFRA was enacted in 1947 and has since been amended numerous times. The primary focus of the Act is the regulation of the manufacture and distribution of pesticides. Unlike most federal environmental laws that either mandate or encourage state cooperation in regulation and enforcement, FIFRA precludes the states from imposing different labeling and packing requirements than mandated by FIFRA and EPA regulations.

FIFRA imposes criminal sanctions for the distribution or sale of pesticides that are not registered with the EPA or pesticides with a composition different from the registration or that are misbranded or adulterated. It also imposes criminal penalties for those who detach, alter, deface, or destroy any labeling required by the Act; and those who refuse to prepare, maintain, or submit records or permit entry, inspections, and copying by EPA officials. The Act also prohibits the use of any registered pesticide in any manner inconsistent with its labeling, falsification of any application for a permit, or to violate any suspension or cancellation order issued with respect to a pesticide (7 U.S.C. §§136j, 136*l*).

Wildlife Protection

Endangered Species Act

The Endangered Species Act was enacted in 1973 to protect plants, animals, fish, and other wildlife in danger of becoming extinct. Unlike most other environmental laws, the EPA is not the agency authorized by the Act to promulgate regulations and enforce its provisions. Instead, the Department of the Interior and the Department of Commerce are the designated agencies.

The Act specifies several types of conduct that, unless an exception applies, are prohibited, including the importation and exportation of endangered species; the capture of endangered species within the United States or on the high seas; the delivery, receipt, transportation, or shipment in interstate or foreign commerce of any endangered species; the sale or offer for sale of endangered species in interstate or foreign commerce; or the violation of any regulation promulgated under the Act or international convention (16 U.S.C. §1538).

The federal laws discussed here are the major environmental laws enacted by Congress. There are, however, many other laws applicable to specific hazardous substances that have been enacted. They are simply too numerous to discuss here. It is important to note that, like those discussed, they impose criminal penalties for knowing violations of their provisions.

The State Environment

Every state has developed its own regulatory scheme. Some of the laws have been enacted in response to federal mandates, and others are based on the specific environmental issues in that state. The states have enacted criminal laws designed to punish those who have harmed the environment and to deter those who might seek to do so in the future.

Issues Affecting Criminal Liability

The Required Mental State for Environmental Crimes

Most environmental criminal statutes require the government to prove that the defendant acted *knowingly*, meaning that the defendant knew that it was engaging in certain conduct. Certain environmental laws also impose liability for those acting with lesser culpable mental states (knowing endangerment; negligence) or with no mental state at all (strict liability).

Knowingly

The Clean Air Act, for example, imposes criminal liability for those who "knowingly" violate various provisions of the Act; or knowingly file false statements, conceal material facts, or fail to file required documents or certifications; or knowingly fail to pay any fee required by the Act (42 U.S.C. §7413). Under this Act as well as other federal environmental laws that require proof that the defendant acted knowingly, the government is not required to prove that the defendant knew the proscribed conduct was prohibited by a specific law or had criminal consequences; it is enough that the government proves that the defendant knowingly engaged in the prohibited conduct (*United States v Weitzenhoff*, 35 F.3d 1275 [9th Cir. 1993]).

Knowing Endangerment

The Clean Air Act, for example, imposes criminal liability for "[a]ny person who negligently releases into the ambient air any hazardous air pollutant" in violation of certain provisions of the act "and who knows at the time that he thereby places another person in imminent danger of death or serious physical injury" (42 U.S.C. §7413[c][4]). This is known as the mental state of knowing endangerment. Basically, it is a combination of the mens rea of negligence pertaining to the act of releasing the pollutant knowingly as it pertains to the consequence of endangering serious physical injury or death.

Negligence

Some environmental laws impose liability upon those who act negligently. For example, the Clean Water Act imposes misdemeanor criminal liability upon those who negligently violate various provisions of the Act, including the violation of permit requirements and the introduction of pollutants or hazardous substances into a sewer system or public sewage treatment facility (33 U.S.C. §1319[c][1]). Under that Act, if a second violation occurs through negligence, the penalty becomes a felony.

Strict Liability

Strict liability means that criminal liability may be imposed without the need to prove that the defendant acted with a criminal state of mind; the mere fact that the violation occurred is sufficient

The Application of General Criminal Statutes

Federal and state prosecutors are not restricted to charging the violation of a specific environmental crime. Often an offender's conduct will also violate criminal statutes that apply to a wide variety of criminal conduct. On the federal side, prosecutors may also charge a defendant with conspiracy, mail fraud, wire fraud, false statements, false claims, or a violation of the Racketeer Influenced and Corrupt Organizations (RICO) Act.

Mail Fraud

The mail fraud statute (18 U.S.C. §1341) imposes criminal liability upon those who engage in a scheme or artifice to defraud or who make false representations and use the mail (including interstate carriers of mail, such as FedEx and UPS) to further the scheme or fraudulent representations.

Wire Fraud

The wire fraud statute (18 U.S.C. §1343) prohibits the same types of fraudulent schemes or promises that are furthered by the use of an interstate or foreign wire communication.

False Statements

The general federal false statement crime (18 U.S.C. §1001) imposes criminal liability upon those who make material false statements to officers, agents, and employees of federal agencies or political bodies performing executive, legislative, or judicial functions. For example, Martha Stewart was charged and convicted of making false statements to agents of the Federal Bureau of Investigation (FBI) and Securities and Exchange Commission (SEC) in connection with an insider trading investigation.

The Racketeer Influenced and Corrupt Organizations (RICO) Act

The RICO statute generally is intended to impose criminal penalties upon those who willfully operate as a criminal enterprise or infiltrate a legitimate enterprise for the purpose of engaging in a pattern of racketeering activity. Although environmental crimes are not included in the list of those crimes that constitute predicate racketeering activity, the crimes of mail fraud, wire fraud, and false statements are listed as predicate racketeering activity (18 U.S.C. §1961).

Conspiracy

The general federal conspiracy law (18 U.S.C. §371) imposes criminal liability upon those who agree to commit violations of federal environmental laws. A conspiracy is separate from a commission of the act. In other words, even though the co-conspirators may not succeed in committing the crime that is the object of the conspiracy, they may be convicted of conspiracy based merely upon the agreement, plus one overt act committed in furtherance of the planned crime. For example, assume that two partners operate an automobile service and repair shop. In the course of the operation, the partners decide to dispose of hazardous petroleum waste by dumping it down the public sewer. One of the partners wheels a drum of the waste toward the street curb with the intention of dumping the waste into the sewer. A neighbor observes the conduct and intervenes to prevent the partner from actually dumping the waste. Under those circumstances, both partners would be liable for conspiracy to violate the Clean Water Act.

The Offenders

Environmental laws impose criminal liability upon individuals and business entities, including corporations. Individuals are liable for the conduct that they, themselves, commit. Corporations, however, are artificial entities that can only engage in criminal conduct through the acts of their employees and agents. Under federal law, corporations may be criminally liable for conduct committed by employees provided that they are acting within the scope of their employment and for the benefit of the corporation. Not all states follow that rule. Those states that have adopted the organizational liability principles of the Model Penal Code impose criminal liability upon corporations only when the conduct of the employees was directed to be done by a high managerial agent or where the employee's conduct was condoned or ratified by a high managerial agent.

In many instances, federal and state prosecutors will seek to hold corporate officers personally liable for the conduct of the corporation even though the officers, themselves, did not engage in the conduct. The theoretical basis for such liability is the responsible corporate officer doctrine.

Responsible Corporate Officer Doctrine

The responsible corporate officer doctrine was developed in criminal cases that arose out of the Pure Food and Drug Act, specifically *United States v Dotterweich* (320 U.S. 277 [1943]) and *United States v Park* (423 U.S. 658 [1975]). In *Dotterweich*, the defendant was the President and General Manager of the Buffalo Pharmacal Company. The company engaged in the business of purchasing drugs from manufacturers and repackaging them under its own label for sale. The Pure Food and Drug Act imposed criminal liability upon "persons" who placed adulterated or misbranded drugs into the stream of interstate commerce. The company mislabeled two drugs that were sold to retailers. Dotterweich did not personally engage in the mislabeling and was unaware that it had occurred. Nevertheless, the Supreme Court upheld his conviction for the violations of the Pure Food and Drug Act because he had a "responsible share" in the process that led to the violations. The Court concluded that Congress properly imposed liability upon the responsible officers who had the opportunity to prevent the violation from occurring rather than imposing the hazard upon unsuspecting and helpless citizens. Similarly, in *Park*, the defendant was the President of Acme Markets, then a national retail food chain with 874 stores and 16 warehouses. The executive offices of Acme were in Philadelphia, PA. An inspection of a warehouse in Baltimore, MD, revealed violations of the Food and Drug Act, and Park was notified of the violations. Park contacted the plant manager of the Baltimore warehouse and directed him to rid the facility of the violations. A second inspection revealed that the violations persisted, and federal prosecutors instituted a criminal action against Acme Markets and Park. Again, the U.S. Supreme Court upheld the conviction of Park. Notably, in both cases, the violations were strict liability crimes, meaning that liability is imposed simply because the violation occurred and also upon the corporate executive simply because he was in a responsible position to prevent the violation from occurring.

Does the responsible corporate officer doctrine apply to environmental crimes that include the means rea of "knowingly?" The following case studies are helpful in considering this issue.

Case Study: Environmental Crime and the Responsible Corporate Officer Doctrine

The three case studies that follow involve criminal prosecutions against a corporate executive for a violation of the RCRA by the corporation's employees. In each instance, the RCRA provision requires proof that the defendant acted "knowingly" and the issue in each case is what proof is necessary to impose liability upon the responsible corporate officer.

United States v MacDonald & Watson Waste Oil Corporation (933 F.2d 35 [1st Cir. 1991])

The Master Chemical Company produced chemicals primarily for use in the shoe industry. In 1982, employees of Master Chemical discovered that water was entering an underground storage tank that contained toluene. Master Chemical emptied the tank and discontinued its use. In 1984, Master Chemical was sold, and the toluene tank was excavated and removed. At that time, a small hole was observed in the tank, and the soil surrounding the tank appeared wet and black and smelled of toluene. McDonald & Watson Waste Oil Company submitted a bid for the excavation, transportation, and disposal of the toluene-contaminated soil.

McDonald & Watson was in the business of transporting and disposing of waste oils and contaminated soil and operated a disposal facility on land in Providence, Rhode Island, known as the Poe Street Lot, that it leased from the Narragansett Improvement Co. (NIC). McDonald & Watson operated the lot under NIC's state RCRA permit that authorized the disposal at the lot of liquid hazardous wastes and soils contaminated with non-hazardous wastes such as petroleum products. Neither NIC nor McDonald & Watson held a permit authorizing the disposal of hazardous waste such as toluene-contaminated soil at that lot. Eugene K. D'Allessandro was the President of McDonald & Watson and the manager of operations at the Poe Street Lot.

The McDonald & Watson bid to remove the contaminated soil was approved, and the contract for removal was negotiated and signed by Frances Slade, a McDonald & Watson employee. Faust Ritarossi, another McDonald & Watson employee, supervised the transportation of the contaminated soil to the Poe Street Lot in ten dump truck loads. A manifest accompanied each truckload bearing a Massachusetts hazardous waste code, and McDonald & Watson employees received an "Authorization to Accept Shipment of Spill Cleanup Material" bearing Slade's name describing the material as toluene. An employee of McDonald & Watson stamped the manifest accepting the material as non-hazardous in Rhode Island. Slade, Ritarossi, and D'Allessandro appealed their convictions for violations of the RCRA. At trial, there was evidence that D'Allessandro participated in the day-to-day management of the company; that he had been warned on other occasions that the company had disposed of toluene-contaminated soil and that it was illegal; but there was no direct evidence that he had knowledge of the particular shipments at issue.

During the trial, the trial judge instructed the jury that the government was required to prove that D'Allessandro was a corporate officer who had the responsibility to supervise the activities in question and that "the officer must have known or believed that the illegal activity of the type alleged occurred." D'Allessandro claimed on appeal that use of the responsible corporate officer doctrine to establish that he acted knowingly was improper. The First Circuit Court of Appeals agreed. It stated, "We agree with D'Allessandro that the jury instructions improperly allowed the jury to find him guilty without finding that he had actual knowledge of the alleged transportation of hazardous waste" on the specific dates involving the Master Chemical soil. The Court concluded that "a mere showing of official responsibility … is not an adequate substitute for direct or circumstantial proof of knowledge."

United States v Self (2 F.3d 1071 [10th Cir. 1993])

In 1981, Steven Self and Steven Miller formed EkoTek, Inc. Self was the President and Miller the Vice President. Self provided most of the capital, and Miller the technical expertise. EkoTek purchased an industrial facility in Salt Lake City, Utah, and began re-refining used oil into marketable products. Both Self and Miller managed EkoTek on a day-to-day basis. The industrial facility operated as an RCRA interim status treatment, storage, and disposal facility before the purchase, and continued to do so under EkoTek.

In 1986, EkoTek began marketing itself as a hazardous waste recycling facility, advertising that it was a licensed facility. EkoTek entered into an agreement with the Southern California Gas Company to transport, burn, and/or dispose of natural gas condensate for the price of $2.50 per gallon. The parties agreed that the condensate was hazardous waste and had to be transported under an RCRA manifest. Shortly thereafter, an EkoTek tank truck driver picked up a shipment of the condensate with instructions to deliver it to the EkoTek facility. As was his routine, the driver stopped at a gas station in Barstow, California. On instructions from Self, the driver's supervisor told him to leave the trailer containing the condensate at the station and return to Los Angeles to pick up some unrelated materials. Self then telephoned the service station manager and instructed him to blend the condensate with gasoline at a 5–10% mixture and add an octane booster. The mixture was then sold as automotive fuel. Self instructed Miller to tell their refinery manager to sign the manifest to indicate the shipment of condensate was received and to falsify EkoTek's operating log accordingly. The manifest was then mailed to the Southern California Gas Company.

EkoTek also started receiving 55-gallon drums of waste material from several sources. Self instructed an employee to store the drums in the south warehouse. When that warehouse became filled, Self told an employee to store drums in the east warehouse. In 1987, the state of Utah issued an RCRA permit, but the permit did not allow for the storage of waste in the east warehouse. Self and Miller discussed the illegal storage. The doors to the east warehouse could be viewed from Self's office, and on several occasions, Self instructed employees to close the door when informed that inspectors would be visiting the facility.

Ultimately, EkoTek went out of business. A broker responsible for arranging for delivery of waste from Avery Label and Reynolds Metals observed drums of hazardous waste and that the hazardous waste labels had been removed from the drums and numbers painted on the drums. The government charged Self with four counts of violating the RCRA and other offenses in connection with the operations of EkoTek. Self appealed his conviction on several grounds. One claim was that, regardless of whether sufficient evidence was presented to show that Self knew that hazardous waste was being stored in the east warehouse, he did not know that the waste from Avery and Reynolds was stored there. The Court disagreed, stating,

> There was direct evidence that Defendant had knowledge of prior illegal storage, and Defendant directed his employee to store hazardous waste in the east warehouse. The jury could infer from Defendant's overseeing of the bills that he knew about the particular waste at issue in count 8. Certainly, the government may "prove a defendant had actual knowledge of a material and operative fact by proving deliberate acts committed by the defendant from which actual knowledge can be logically inferred." (*United States v Uresti-Hernandez*, 968 F.2d 1042, 1046 [10th Cir. 1992])

United States v Johnson & Towers, Inc. (741 F.2d 662 [3rd Cir. 1984])

Johnson & Towers, Inc. operated a facility for the repair and overhaul of large motor vehicles in Mount Laurel, New Jersey. In the course of its operations, the company used degreasers and other industrial chemicals that were classified as hazardous waste under the RCRA. Waste chemicals from its cleaning operations were drained into a holding tank, and when the tank became full, employees pumped the waste into a trench that flowed from its property into Parker's Creek, a tributary of the Delaware River. The RCRA required a permit for the disposal of such waste, and the company did not apply for nor did the EPA issue such a permit.

The government charged Johnson & Towers, as well as Jack Hopkins, a foreman, and Peter Angel, the service manager in the trucking department, with three counts of violating the RCRA as well as other charges. Hopkins and Angel moved to dismiss the RCRA charges, claiming that liability extends only to owners and operators, and not to corporate managers who were not in a position to secure the required permit. The Court disagreed, concluding that the responsible corporate officer doctrine could be applied to regulatory violations of the RCRA intended to protect public health and that "all of the elements of that offense must be shown to have been knowing, but that such knowledge may be inferred by the jury as to those individuals who hold the requisite responsible positions with the corporate defendant."

In each case, the issue concerned proof that a responsible corporate officer acted "knowingly." The case of *Johnson & Towers* seems to suggest that such knowledge may be inferred from the mere fact of the defendant's status as a responsible corporate officer. In *McDonald & Watson*, the Court clearly rejected that view, concluding that actual knowledge must be proven but that such knowledge may be established by circumstantial evidence, one of the circumstances (but not the only one) being the officer's responsibility for the actions of the corporate employees. That also is the view stated by the Court in *Self*. The U.S. Supreme Court has yet to decide the issue.

Summary

During the 1970s, the federal government enacted several major laws designed to provide comprehensive regulation of the environment. In addition to the creation of the Environmental Protection Agency and the delegation to the EPA of significant regulatory authority, Congress mandated the exercise of regulatory and enforcement powers by the states and Indian tribes. The federal government has established a comprehensive environmental policy, and Congress is fairly responsive to emerging environmental threats by the enactment of legislation specifically designed to protect the environment.

The overwhelming mens rea set forth in environmental laws is "knowingly." The Courts have struggled somewhat with the meaning of "knowingly" and as well with the application of the responsible corporate officer doctrine to that mens rea.

Discussion Questions

1) The Courts appear in general agreement that corporate executives can be held liable for knowing violations of the RCRA even though they personally did not engage in the particular activity at issue. What is not so clear is the extent of knowledge that the government must prove. What do you think? When companies engage in the transportation and disposal of hazardous waste, shouldn't the corporate executives face criminal liability for the acts of the corporation, especially when the conduct presents a public health and safety issue? On the other hand, is it appropriate to impose liability for acting "knowingly" when someone had no actual knowledge of the illegal conduct? Think about newsworthy incidents that caused major harm to the environment. Could they have been prevented if imposition of liability merely because of the position of individuals as executives served as a deterrent?

2) The federal government is clearly the dominant force in establishing environmental policies, standards, and enforcement regimes. Is this appropriate when certain environmental dangers are local or regional in nature? Shouldn't the states have the authority to impose environmental safeguards even though they conflict or exceed federal standards? What counterargument would you raise to that suggestion?

Bibliography

DiTata, B. (1996). Proof of knowledge under RCRA and use of the responsible corporate officer doctrine. *Fordham Environmental Law Review*, 7, 795.

Hustis, B., & Gotanda, J. (1994). The responsible corporate officer: Designated felon or legal fiction? *Loyola University Chicago Law Journal*, 25, 169.

St.John, D., Brannan, C., Beiderwieden, H., Fountain, J., Larson, M., Lydic, J., & Stegman, L. (2020). *Environmental crimes*, Thirty-fifth annual survey of white collar crime. *American Criminal Law Review*, 57, 657.

Thirty-fourth annual survey of white collar crime. *American Criminal Law Review*, 56, 807 (Summer 2019).

United States Code (U.S.C.) (2020). Title 7, Sections 136j, 136l; Title 15, Sections 2614, 2689; Title 16, Section 1538; Title 18, Sections 371, 1001, 1341, 1343, 1961; Title 33, Sections 1251, 1319, 1362; Title 42, Sections 6903, 6928, 7401, 7413, 9603, 9604, 9607, 9612. Available at https://uscode.house.gov/.

Chapter 5

The Driving Forces of Environmental Criminality

As a society, we have continually evolved technologically. This evolution has palpable benefits, but as discussed in preceding chapters, the evolution has a "dark side" as it has given rise to burgeoning volumes of manufacturing wastes, much of it harmful to health in many ways. Along with our enhanced understanding of the extent to which exposure to these wastes can prove dangerous, we have sought advanced methods of minimizing and neutralizing the deleterious effects of these substances. These methods are now widely available and give us the ability to conquer the harmful qualities of this outgrowth of technological progress, affording us varied means to render these damaging materials innocuous. So, why are there those who would avoid the use of these methods? What are the driving forces that are responsible for transforming legitimate, honest individuals into environmental criminals? This chapter will address these two important questions.

In her article *Is Illegal Dumping of Hazardous Waste a Viable Business Strategy?*, Dawn DeVroom (September 24, 2013) points out that with the emergence of the green movement over the last 20 years, many Americans have been drawn into debates over global warming, illegal dumping, and the effects of toxic waste on the environment. However, she stresses that a comprehensive exploration of these issues has not necessarily had an effect on the illegal dumping of hazardous wastes. She cites as the primary "motivations" for those who would dispose of illegally, (1) the avoidance of disposal fees, (2) companies with an untrained and understaffed workforce not understanding proper waste management procedures, (3) avoidance of the time and effort required for proper disposal, and (4) the uncertainty that some firms may have about what constitutes hazardous waste. DeVroom aptly notes that cases of illegal disposal can be relatively small and may seem to be benign to the offenders or they may be quite large and implicate major corporations that are seemingly conducting these activities routinely without regard for public health and environment (DeVroom, 2013; O'Donnell, 2013).

She gives as an example of the first type a case in which an Alabama school of math and science was fined $10,000 by the Alabama Department of Environmental Management after school officials buried 24 computer monitors instead of following proper disposal guidelines. Here, state environmental officials admonished the school for their disregard that the monitors contained potentially hazardous waste including lead and silver. DeVroom then shifts her attention to the other end of the environmental offender spectrum in the form of big-box retailers like Wal-Mart agreeing to pay $82 million in fines for improperly disposing of insecticides, bleach, and fertilizer into municipal trash dumpsters and sewer systems in California and Missouri. DeVroom hastens to state that Wal-Mart is not alone, citing violations by such large companies as Walgreens, Costco

Wholesale, CVS, and Target. Each of these cases ended in the generation of fines in the millions to tens of millions of dollars (DeVroom, 2013).

Does DeVroom's premise have merit? Are the four general areas she proposes a valid representation of driving forces of environmental criminality? And what of the title of her piece? Does it hint at a position that the greatest danger lies with those who would make a rational choice of environmental criminality as "good business sense?"

The Offenders: Why They Do It

In the preceding summary of DeVroom's analysis, her model of driving forces for environmental criminality can really be broken down into a single dichotomy: those who know what they're doing and do it intentionally and those who may not know what they are doing and do it without malice. Rebovich's multi-year empirical study of environmental crime in the northeastern part of the United States (The Northeast Hazardous Waste Research Project) was able to develop three general categories of environmental offenders based upon the crimes they committed and how they committed them. The first category was termed *situational environmental crime*. These are crimes that are conducted by the environmental offender who may conduct the offense intermittently without any intention of committing additional offenses as part of a routine. An example of this could be a car owner who decides he will save some money on changing the oil in his car by not having the job done by a mechanic, but by doing it himself. The driving force here may be simply that this individual is trying to cut costs in a particular month for financial reasons and changing his own oil may be one of a list of actions he'll take and then return to previous practices in succeeding months. Technically, the dumping of the pollutants contained in the used crankcase oil disposed into a ditch or a sewer may not amount to much by itself. But, nevertheless, it is an offense against the environment. The Northeast study described a more serious example of a situational offense in the form of a fish packing plant located near a major waterway. The situation was that the plant was in the process of shutting down and the owners would be left with refrigerant waste that would be required to be professionally treated before disposal. By chance, a major storm hit the area where the plant was located. The owners disposed of the waste contending that the facility had been damaged by the storm releasing the refrigerants into the waterway. Through careful investigation, it was determined that this explanation was a ruse, and the offenders were punished. Once again, these were offenders who demonstrated no sign of conducting such illegal activities prior to this confluence of financial need and natural disaster. Clearly, they were environmental offenders, but not ones who would, necessarily, demonstrate any signs of repeated criminal activity in this vein (Rebovich, 2015).

The second form of offender referred to in the Northeast study was those who committed environmental crimes in a *routine* fashion (i.e., *routine offenders, routine environmental crime*). These were cases in which the offenders were operating or working for legitimate entities that provided legitimate products or services. All in all, these were organizations and agencies that would go through everyday work activities in a professional and honorable manner. However, these routine offenders were found in some way to subject a part of their operations to routine violations of environmental regulations and environmental laws for either cutting losses or simply sidestepping necessary activities. These activities did not represent the core of the entities but were a meaningful part of general operations. In this sample of cases studied in the Northeast study, an example of a routine environmental offender was represented by a well-known public university. The university's medical school would routinely dispose of the carcasses of experimental animals without proper treatment, representing significant biological threats. Another example of routine environmental criminality that was not part of the Northeast study sample exists in the form of past practices taken by Royal Caribbean Cruise lines with regard to the disposal of wastes generated on vacation cruises. The untreated disposal of these wastes at sea was detected by regulatory agencies and the U.S. Coast Guard in single incidents in one location. What on the surface seem like isolated activity by the cruise line proved more nefarious as surveillance continued on other ships located in different parts of the globe. The bottom line was that Royal Caribbean Cruise lines were involved in routine illegal disposal that led to one of the highest fines paid by a corporation for these actions (Martinson, 1999).

The primary difference of routine environmental criminality in the last category, *entrepreneurial environmental crime*, is that with *entrepreneurial environmental offenders* the individuals and organizations are internally dedicated to not just cutting costs but making a profit off of continuous violation of environmental regulations and criminal laws. The Northeast study pointed to offenders in the hazardous waste treatment, storage, and disposal industry in New Jersey as examples of entrepreneurial environmental offenders. While the owners and operators of such facilities may have started out adhering to ethical practices and procedures, a pronounced desire for obtaining greater profits got the best of them. In some cases, the combination of a quest for optimized profits along with the emergence of stiff competition from other local facilities was responsible for operators to venture beyond the mere cutting of corners. These operators would actively devise schemes to criminally dispose of hazardous wastes it accepted to be legitimately treated, fouling land and water and defrauding waste generators paying top dollar for conventional treatment of their manufacturing wastes. Such facility operators would typically begin their criminal activities by withholding treatment of a small percentage of incoming hazardous wastes (e.g., 15% of total waste volume) and simply dumping the wastes, sometimes on facility land or in abutting waterways. As time went on, without detection of their violations, these facility operators became criminally bolder, increasing the volume of untreated waste to be dumped. In some cases, facilities converted their operation into a complete guise of a facility in which 100% of incoming wastes were surreptitiously and criminally disposed of. In effect, these criminal entrepreneurs were bringing in the wastes in the front door and dumping them out the back. These are environmental offenders who had "graduated" to full-fledged career criminals (Rebovich, 2015).

The "Calculating" Criminal

In her depiction of environmental criminals, DeVroom posits that some of these offenders commit their crimes either to avoid disposal fees or to avoid the time and effort required for proper disposal. These are the criminals who commit their crimes purposely, not by accident or because of a lack of knowledge. A certain amount of pre-planning has to go into the process of determining how to commit a crime and how to escape detection. When considering driving forces that play a role in this thought process, one can turn to criminologist Donald Cressey for a logical explanation as to what the driving forces are that can lead to the commission of environmental crime. Cressey's rational choice theory explains that there is a triangle of key factors that, taken together, will result in a criminal act (Cressey, 1972).

Those three factors are (1) pressure, (2) opportunity, and (3) rationalization. As explained by Cressey, pressure can be represented by a financial need view by the offender. With regard to crime in general, this can take the form of finances needed to feed some type of addiction like substance addiction. In cases of embezzlement, this pressure can take the form of a desire to meet certain financial goals or take an individual out of debt. Cressey points out that the pressure can be artificial yet perceived to be real by the offender. In either case, the presence of this pressure is critical in understanding the rational thought process of criminal offenders. In the case of environmental criminality, this pressure may exist as a result of a perception that a business entity can ill afford to withstand the perceived burden of incurring costs for the legitimate treatment of manufacturing wastes to legally dispose of them. The perception of pressure could also entail the perception that the business entity would not be able to adequately compete with business competition if the costs for the proper treatment of manufacturing wastes are routinely paid out. This was revealed as a major driving force for criminality in the Northeast study, particularly with regard to small waste generators (SWGs), small businesses that generated hazardous wastes but found it difficult to make ends meet with the perceived high costs inherent in properly treating generated wastes. In some cases, owners expressed that they believed their businesses would "go under" if demand for their services did not increase while the cost of proper treatment of generated wastes would rise over time.

Opportunity for Environmental Offenders: The California Manifest Tracking System

The second factor that Cressey explains as being crucial in the decision-making to commit a criminal offense is opportunity. Here, Cressey is referring to the calculation made by the offender regarding the chances that the offender would be able to commit the offense without being detected and avoid the risk of punishment for his deeds. If the offender feels that the odds are in his favor to avoid detection, for whatever reason, it tips the calculation in favor of committing the offense. For environmental crime, the opportunity for avoiding detection can come in the form of the natural environment in which the offender operates. "Midnight dumping," the practice of offloading hazardous wastes and garbage in the wee hours of the morning, can take place in any environment. In the past, the areas ripe for illegal disposal were vacant land and secluded wooded areas. More recently, offenders have become more brazen and have disposed wastes in public parks and even on the streets of cities. Commonly, offenders have explored surrounding areas for places they perceive they could dispose in without being seen. Such areas have included abandoned coal mines, abandoned gas stations, waterways, and even farmlands. An additional factor to consider in calculating opportunity has to do with concocting a plan to avoid detection by inspectors and regulators. For the Northeast study, this included a careful assessment of regulator inspection patterns and the timing and predictability of inspection visits. Government manifest tracking systems are in place to help track hazardous wastes from "cradle to grave" requiring the completion of manifests by waste generators, waste transporters, and waste traders to ensure that wastes are treated properly. The problem is that these systems are not foolproof and that knowledge is not lost on calculating environmental offenders.

The state of California stands as a case in point in which offenders have taken advantage of the opportunity presented by a tracking system that has left much to be desired. A *Los Angeles Times* investigation in 2013 discovered that despite California having some of the strictest cradle-to-grave tracking systems in the country, the state has chronically failed to keep track of thousands of tons of dangerous chemicals and toxic metals transported for disposal on the state's roads and railways. In California, businesses that generate waste are required to report every load they transport, and disposal and treatment facilities are required to record the wastes' arrival. California's Department of Toxic Substances Control is responsible for tracking wastes to make sure they are not disposed of illegally. The newspaper's investigation found gaping holes in the department's database, regulators limiting the use of available information, and the inability to automatically flag potential problems. It was found that California's Department of Toxic Substances Control was not able to account for 174,000 tons of hazardous waste material shipped for disposal from the years 2008 through 2013. While the existing database showed that the wastes were shipped, the destinations of the wastes remain murky. The wastes included more than 20,000 tons of lead, 520 tons of benzene, and 355 tons of methyl ethyl ketone. This substance is a flammable solvent and is often referred to in the waste hauling industry as "methyl ethyl death." One of the inherent problems with the system has been that, for years, it has not been computerized, instead relying on paper signatures and mail delivery (Garrison, Poston and Christensen, 2013).

On the surface, the manifest system in California seems effective enough. Businesses that are shipping toxic wastes generated by them are required to complete a manifest and forward a copy to a post office box in Sacramento. The manifest is made up of an original form with five carbonless copies. The bottom copy goes to the waste generator, and the others go to the haulers. The inherent problem is that the bottom copy has the faintest writing impressions. The easiest to read copies on top go to disposal and treatment facilities. These copies are used to ensure that whatever was delivered matches what was shipped. The facility accepts the load if there is a match and fills out and sends copies of the manifests to Sacramento. Theoretically, this cradle-to-grave system should ensure that there is no opportunity for offenders to take advantage of the process. However, illegible paper forms present that opportunity to cunning offenders in California. In many cases, it was found that state contractors who are responsible for data entry did not enter information for many shipments because they cannot read the information on the generator's copy. This has resulted in the inability to effectively verify that wastes received by treatment and disposal facilities are the same wastes and volumes that were shipped. The *Los Angeles Times* reported that disposal manifests that document that shipments are received and processed appropriately were missing for a percentage of cases that equated to 174,000 tons of missing waste (Garrison, Poston and Christensen, 2013).

It was clear from the newspaper's further investigation that a number of entities have capitalized on the holes in California's tracking system. By directly contacting major disposal facilities, it became apparent that significant amounts of waste never got to their intended destination. One electroplating company reported sending close to a ton of waste to one facility in Arizona that had no record that the wastes were ever received and, in the meantime, the plating firm went out of business. A number of facilities declined the newspaper's request for information on thousands of missing loads. Absent an effective tracking system, it is believed that California offenders took the opportunity to dump wastes down drains, on roadsides, or in public landfills. In one case a load of wastes documented in a manifest was found in 55-gallon drums in Santa Fe Springs, California, located in a small warehouse across from a residential neighborhood. The most egregious of the offenses occurred in the small desert town of Mecca, California, on land owned by the Cabazon

Band of Mission Indians. There, a soil-recycling plant had been accepting over 160,000 tons of metal sludge, empty pesticide cans, and other wastes from construction sites in Los Angeles over a two-year period. The irony was that project managers at the state's regulatory agency had authorized the shipments and the totality of the mistake had not been understood until odors from the facility had made children in a nearby school ill and had prompted involvement by U.S. Senator Barbara Boxer. A search of the database found records of all the shipments. The problem was the state agency had failed to monitor its own database (Garrison, Poston and Christensen, 2013).

In Cressey's rational choice theory, the larger the opportunity presented to potential criminals, the greater the chance criminals will take advantage of that opportunity. From a prevention standpoint, the best strategy that enforcement officials can take is to do everything possible to lessen that opportunity. In the case of the California tracking system, the U.S. EPA has taken steps to develop a national electronic manifest system to upgrade the paper-based system of tracking hazardous wastes to an electronic one. The action authorizes the use of "e-manifests" to track hazardous wastes under the Resource Conservation and Recovery Act, allowing the current process, which requires paper forms, to be streamlined. Cressey has posited that calculating criminals can be very dogged in continually assessing opportunities and subsequently adjusting methods to meet new opportunities. Such has been the case with environmental criminals over time.

"Rationalizing Away" Environmental Criminality

Rationalization is the third point of Donald Cressey's rational choice triangle. A common definition of rationalization is to attempt to explain or justify one's own or another's behavior or attitude with logical, plausible reasons, even if these are not true or appropriate. In cases of embezzlement, it is not uncommon for admitted embezzlers to confess that they were only "borrowing" from their employer with an intent to pay back their "withdrawals" at a later date when they are more financially stable. Another common justification given by embezzlers is that they feel somehow their employer has not treated them fairly particularly with regard to lack of promotion or lack of raises. These individuals will use these reasons to justify the commission of embezzlement and to avoid a sense of criminal guilt. As it would apply to environmental criminality, the act of rationalization would manifest itself in the attempt by the environmental offender to somehow reconcile their criminal deeds with the belief that in some way the act is either justified or excusable.

In the Northeast Hazardous Waste Research Project, it was found that rationalization took place in several forms. As expressed by representatives of several small businesses, the rationalization used for illegally disposing manufactured wastes was that the cost of the legitimate treatment and disposal of wastes proved prohibitive to the business and would result in actions like employee layoffs. These offenders would use this rationalization as an expression of a form of "altruism" toward their subordinates and convince themselves that they were committing the acts to help their employees and their families. Another form of rationalization among small waste generators was that treatment and disposal facilities were found to be scarce in their region requiring the generators to incur extra cost for the shipping of these wastes to facilities far from the location of the waste generators. In many of these cases, the waste generators were telling the truth, in that due to the difficulty of siting new facilities in their region, because of community resistance to the physical siting of these facilities near residential areas and schools, these waste generators were shouldering an increased burden of long-term costs if they were to obey the law. As one business owner lamented in investigative interview records, "I felt like I was between a rock and a hard place." In short, many of these generators believed that government had "overregulated" with increased regulations that they could not feasibly adhere to and continue to run their businesses. Hence, this feeling of being treated unfairly translated into a conclusion that the illegal disposal of hazardous wastes was forgivable in their minds and gave them a sense that they were not, in fact, *true criminals* (Rebovich, 2015).

Another form of rationalization found to be used by environmental offenders is that the practice of the illegal disposal of hazardous materials was not nearly as dangerous as state and government officials would have us believe. This attitude is perfectly captured in the 1986 TV movie *A Deadly Business*. The film is based on the true story of "whistleblower" Harold Kaufman, an ex-convict turned FBI informer who exposed corruption and involvement of organized crime in the handling of toxic wastes. In the film, Charles Macaluso, Kaufman's boss, is sent to prison on Kaufman's testimony. A telling scene in the film is when Kaufman (played by actor Alan Arkin) questions Macaluso on their methods of illegal dumping. The response of Macaluso is, "The Earth is very forgiving" (Taft Entertainment Television, 1986). While it is unclear if this attitude was genuine, The Northeast Hazardous Waste Research Project found that at least some of those responsible for actually disposing of the wastes had convinced themselves that the material they handled could not be "that harmful" if they, themselves, did not experience any signs of health problems. The following is an excerpt from an interview of a "yard worker" from a waste disposal facility.

> When we were getting rid of the wastes, it would take so long to run things through the little hoses and whatnot, and we were burning large quantities of wooden pallets anyway. It was considered to be more expedient to just throw the bottles of chemicals on top of the wood fire whereupon they would usually break and if they didn't break, the heat would explode them and they would burn. And things that didn't burn, like I could throw five or six bottles of sulfuric acid on there and it would evaporate them. Uh, I remember several times some other employees forbid to do that when they were around because they couldn't take the fumes. I had gotten used to it because when burning in the boiler run, the boiler room would fill up with smoke all the time and it had just gotten to the point where it didn't affect me at all. I was totally fine and didn't feel that there was any danger to me at all.

This belief that close exposure to hazardous waste and the breathing of gases emitted from the burning of the wastes was not harmful was demonstrated in a number of interviews of those instructed by their superiors to illegally dispose of wastes. Underlying this rationalization was the understanding that if they had refused to comply with the instructions for illegal disposal, it could mean the loss of their jobs. This was expressed by many who were interviewed in the investigation of cases studied for the Northeast Hazardous Waste Research Project. In many criminal treatment and disposal facilities, managers were careful to weed out those who would not comply with instructions to illegally dispose. Investigators would subsequently turn to those who were fired for not complying with criminal instructions to serve as witnesses against the managers of the company's in resulting criminal cases.

Rationalization is not confined to those committing the criminal act when it comes to environmental crime. Figures who operate outside of the actual criminal act may intentionally or unintentionally play a part in the facilitation of environmental crime. In the Northeast Hazardous

Waste Crime Research Project, it was found that "outsiders" to the criminal act played important roles in offense success. Some former law enforcement officials who worked for state environmental protection agencies and the office of state attorneys general were found to quit their jobs and export their knowledge of the regulation system (and how to get around it) by taking on jobs with those they had investigated in the past. In these cases, a larger paycheck could prove to be the key factor in rationalizing a lowering of ethical standards. Likewise, chemists from laboratories that would be hired for analyzing samples taken from waste dump sites to determine the level of substance toxicity could be found to "advertise" liberal test reading standards, thus sending the message to defense attorneys of the accused to partake in their services as a way of supplying ammunition for defense arguments. These chemists were careful to not step over the line into criminality. Their rationalization was that they were not breaking the law per se, just stretching the truth (Rebovich, 2015).

Sometimes, an unwitting force for rationalizing environmental criminality can be the local community itself. In the book *Toms River: A Story of Science and Salvation*, author Dan Fagin relates the story of the Swiss pharmaceutical firm Ciba-Geigy and the effect its operations had on the town of Toms River in Ocean County, New Jersey. In 1952, Ciba opened a dye manufacturing plant called the Toms River Chemical Corporation. Workers at the chemical plant complained of deleterious effects experienced working there. Ciba was located not far from the Atlantic Ocean coast. The company constructed a pipeline to the ocean through which hazardous wastes from their manufacturing process were dumped into the Atlantic Ocean off the coast of nearby Ortley Beach. The disposal of these wastes through the pipeline continued for over 30 years. An official from Ciba-Geigy reported that the waste consisted of 99% water with a small amount of salt (whether this was rationalization or not is debatable). The substances turned out to be extremely toxic. What is not debatable is what drew the company to Toms River in the first place. Despite a checkered past, the company was welcomed with open arms in the beginning. It was known that the company had been involved in environmental troubles in its original home site of Basel, Switzerland, and at later sites in Cincinnati, Ohio, and in Rhode Island. Yet, these issues were put aside by community leaders. As explained by Dan Fagin:

> Toms River was really a town like any other town. It was a sleepy town down at the Jersey Shore. The economy was somewhat moribund. So when one of the Big Swiss chemical companies, Ciba, started looking for a place to relocate ... Basically everywhere they went they eventually became a most unwelcome neighbor. But when they came to Toms River, initially everyone was thrilled, because this was going to be, or it was ultimately, one of the largest dye manufacturing plants in the country, one of the larger ones in the world. And, that it eventually expanded from dyes into plastics and other chemical products. And for years, it was the most important employer in Ocean County. And it wasn't just that there were a lot of jobs, they were well-paying jobs, good blue-collar jobs. So, in some ways, it was a real agent for social mobility, as long as people didn't think too hard about the long-term consequences. There was a lot of short-term thinking versus coming to grips with the long-term. (Fagin, 2013)

Since the 1990s, families that were affected by cancer linked to pollution have been pursuing a class action suit against Ciba-Geigy. In 1992, Ciba-Geigy agreed to pay New Jersey $62 million

Figure 5.1 Ciba-Geigy Toms River site cleanup operations. https://cumulis.epa.gov/supercpad/cursites/csitinfo.cfm?id=0200078

for illegal waste dumping. After all was said and done, the Toms River Ciba-Geigy plant was no more and 3,000 workers were left without jobs. So, the natural quest for invigorating a declining local economy can sometimes lead to a community "group rationalization" that may dismiss obvious red flags that would open the door for environmental disaster. Toms River, New Jersey, is not alone in this type of group thought process as other towns throughout the United States have followed a similar path. It is just that it happens to be a case that had wide-reaching harmful effects to the citizenry and the community in general (Figure 5.1).

The "Misunderstanding" of Hazardous Waste and Its Proper Management

DeVroom concludes her description of forces integral to the commission of environmental crime by referring to a general sense of unawareness of what constitutes hazardous materials and how they should be properly managed. This does not necessarily constitute a willful ignoring of environmental regulations and laws, but a lack of understanding of the damaging characteristics of hazardous wastes and the handling of these wastes. Turning back to the Northeast Hazardous Waste Research Project, a review of archived investigative interviews of offenders uncovered clear examples of cases in which those responsible for handling hazardous wastes were not cognizant of either the imminent danger of mishandling these materials or the specific manner in which the wastes should properly be handled and disposed of. Case files and interviews established that many of the owners and managers of waste generating and waste disposal entities were typically college educated and had at least a serviceable understanding of the qualities of the substances handled and how they should be properly handled. Unfortunately, that information was not always effectively transmitted to subordinates in those entities; those individuals responsible for direct contact with the substances. In some cases this was an oversight, but in others it was clear that it was an intentional act to prevent these individuals from fully understanding the makeup of the substances. In one case, a worker explained the following under questioning by an investigator:

Investigator: Do you have any scientific background?
Employee: One year of chemistry, one year of biological science curriculum studies.
Investigator: Did you know the chemicals that you were burning from your scientific background or did you learn them through some other means?
Employee: Mr. _____ bought me a book. I believe it was, I forget what it was called, but it's simple and easy ways to dispose of chemicals. But we never followed that, per se. We never got around to it. He's too busy to get rid of the rest of the stuff. I recognize things like sulfuric acid and I knew not to do. I knew to pour sulfuric acid into water and not water into sulfuric acid and stuff like that. But first-year high school chemistry only takes you so far.

In one case, a mid-level manager at a treatment facility claimed to investigators that his job at a facility placed him in a "handicapped situation." He explained that he was never given proper equipment to protect himself from the wastes he handled. He added that he felt this was not unusual because he knew that their competitors did not have adequate equipment either, overlapping an acquiescence to a lack of proper protection with a rationalization that this was part of the business. This particular manager was responsible for screening waste loads for high flammability. From a technological perspective, he understood the proper and safe way to test flammability. He knew that the proper way to do this would be to use a Pensky–Martens closed-cup flash-point test; a brass test cup is filled with a test specimen and fitted with a cover. The sample is heated and stirred at specified rates depending on the material that is being tested. An ignition source is then directed into the cup at regular intervals with simultaneous interruption of stirring until a flash that spreads throughout the inside of the cup is seen. The corresponding temperature is its flash point. The manager reported that this typical protocol was not followed on his job. As he explained:

Apparently they couldn't or didn't want to spend the money. _____ said it costs $800 for a tag or a Pensky–Martens tester. So they used a jar. Every load was screened this way, by lighting a match over the jar to see if it flared for flash. They rejected shipments that way. There were no formalized reports on this.

In other cases, truckers responsible for hauling waste loads to landfills were purposely kept from understanding the toxic qualities of the substances they were transporting to landfills that were not equipped to accept those types of wastes. In one situation the hauler was told by the waste generator that he was hauling wastes that were harmless when they actually contained paint thinner wastes that were flammable. On one occasion the hauler was approached by the landfill operator upon entering the landfill. The landfill operator examined the wastes and segregated them and told the hauler to take them back. According to the trucker:

> The supervisor asked me what happened; where the drums came from. I told him and he said be careful, don't take any more drums like that anymore. When I went to pick up the drums the next day, the landfill employee told me it was flammable stuff which I didn't know at the time I picked it up. We threw the drums into my truck and there was some stuff left in his hopper of the bucket loader there. We threw a match and it went up in flames.

In this research study, there was case after case of situations in which the individuals handling dangerous substances were clueless to the extent of danger they were subjecting themselves to by handling those wastes recklessly. Above them, there were often supervisors who did understand the dangerous nature of the mishandling of the substances. However, because these individuals did not come in close proximity to the substances and were more interested in enhancing profits, they turned a blind eye to the plight of their subordinates. They were content to keep their workers in a state of "blissful ignorance" while they subjected them to the greatest of dangers. In one case, a misguided worker went so far as to characterize the dangerous burning of these wastes as a form of entertainment at his own unwitting risk.

> I was instructed to pour some acid out onto the ground along with ether out onto the ground and let it evaporate, and they set up a big barrel where I was to just pour the ether in and other things that would evaporate quickly. We poured the stuff into a barrel and let them evaporate, pouring in a little water now and then. For some reason he said we were not supposed to just let it sit there. Add water or something. But it ate through the bottom of the barrel and it all went out onto the ground. It was my duty to take all the other barrels in one form or another and just throw them on the fire and let the fire ignite them and blow them apart and burn what was inside them. These were fun fires to watch. Several things, I hadn't the foggiest idea what they were but I was instructed to personally take them out and pour them on the ground. There were a lot of colorful explosions.

Discussion

Several factors that contribute to the commission of crimes against the environment are presented at the beginning of this chapter. These factors are companies with untrained and understaffed workforces not understanding proper waste management procedures; the uncertainty that some firms may have about what constitutes a hazardous waste; the avoidance of disposal fees; and avoidance of the time and effort required for proper disposal. Examples of the first two factors are borne out in criminal case examples in which handlers of hazardous waste substances were not provided with knowledge of the potentially dangerous effects of the substances they were exposed to, nor were they furnished adequate information on the proper manner to legally dispose of chemical wastes handled. The absence of this information is a contributing element to not only endanger the environment but also endanger the health of the waste handlers themselves. Withholding this information can be a convenient tool of those engineering illegal acts of pollution that is explained by considering the latter two contributing factors, avoidance of disposal fees and the work required for legal disposal.

The mindset of those who would commit environmental crimes to avoid the costs and effort that goes into properly treating harmful substances before disposal can be best explained through

Cressey's triad of pressure, opportunity, and rationalization. Businesses that produce hazardous waste are compelled to adhere to strict regulations and laws and how they dispose of these wastes. Naturally, there is a monetary cost for the expected treatment of these wastes to ensure that they are legally disposed of. Some businesses, especially small ones, consider these costs to be unaffordable, prompting business operators to cut costs by disposing illegally. Other business owners, haulers, and operators of treatment and disposal facilities feel a sense of pressure to keep up with their competitors or are simply profit motivated and, consequently, turned to illegal means of getting rid of the wastes. Opportunities to commit these acts without being detected are prevalent. There are many natural characteristics of the environment that supply ready-made "final resting places" for large volumes of chemical wastes. In addition, shortcomings in waste tracking and regulatory systems practically "invite" waste generators and handlers to dispose illegally. Rationalization plays a role in the commission of these crimes. Managers and lower-level employees alike find ways to rationalize their acts by convincing themselves that their acts are not as bad as they seem or that circumstances out of their control induce them to take the illegal disposal path. The end result is the looming menace of environmental pollution.

Discussion Questions

1) This chapter presents three categories of environmental offenders that are based upon the offenders' patterns of crime commission. Explain those three categories. What aspects of each category make the categories unique? Which category of offender do you believe is the greatest threat to the environment and why?
2) According to the chapter, ample opportunity to commit environmental crimes and avoid detection can prove to be enticing to those considering the commission of these crimes. In what ways do you believe society can limit those opportunities to prevent environmental crime? What are the obstacles that would have to be overcome to limit opportunities to commit environmental crime? How do you envision the level of opportunities to commit these crimes in the future? Why?

Case Study: An Environmental Crime Enforcer's Perspective on New Jersey Organized Crime and Environmental Crime in the 1980s

Environmental crime in New Jersey reached its peak in the 1980s. Much of it was found to be committed by small waste generators (SWGs) who committed the offenses individually or within small groups. However, a portion of these offenses did involve organized crime in the hazardous waste disposal industry. One of the most thorough and incisive assessments of this phenomenon was presented before the U.S. Senate Committee on Governmental Affairs Senate Subcommittee on Investigations on February 1, 1984. The following is a summarization of testimony given by then Deputy Attorney General and Chief of the New Jersey Division of Criminal Justice's Environmental Prosecution Section, Steven J. Madonna. At the time of the testimony, DAG Madonna was responsible for the investigation and prosecution of anti-competitive and racketeering activities in the solid waste industry and investigation and the prosecution of the illegal transportation and disposal of hazardous waste (Madonna, 1984).

Mr. Madonna began his testimony by presenting a simple explanation of what hazardous waste is and how it is produced. He painted the picture of hazardous waste as the presence being ubiquitous, generated in villages, towns, and cities throughout the United States. He went on to say that although the public's general perception of hazardous waste is associated with the production of chemicals pharmaceuticals and petroleum, they can also be the byproduct of routine processes like dry cleaning, auto body repair, hospital operations, and farming. Methods of legitimate transportation and treatment of hazardous wastes were explained. Mr. Madonna, in the early part of his testimony, stated that he believed that as much as 90% of the hazardous waste generated in the United States was being disposed of improperly. At the time, Madonna attributed this alleged high percentage of improperly disposed hazardous waste to a generally deficient level of effective criminal enforcement activity involving hazardous waste cases. He characterized environmental crime enforcers on *all* government levels as being woefully unaware, indifferent, or ill equipped to investigate and prosecute typical hazardous waste offenses. Madonna also added that exacerbating the problem was a lack of awareness to the extent to which public health and the environment were adversely affected by the illegal disposal of hazardous waste. He was careful to present this information in a context to further explore how great profits can be made through the infiltration of organized crime into the hazardous waste disposal industry (Madonna, 1984).

In the testimony, it was contended that organized crime had the capability to infiltrate and take over any type of business, including a generator of hazardous wastes. The example given was a situation in which a hazardous waste generating company is taken over by a syndicate crime group, leading to the illegal disposal of hazardous waste generated. It was pointed out that the greater danger of organized crime infiltration would reside in the latter stages of the disposal process. The rationale underlying this statement was that a generator-by-generator takeover would amount to a takeover of the entire commercial/industrial complex in the United States. Madonna testified that he believed it would be virtually impossible to impact the hazardous waste disposal industry through this type of approach. He explained that for organized crime to achieve a successful takeover of an industry, it seeks to find some "pressure point" or "focal point" crucial to the successful operation of the industry. Taking treatment and disposal components as an example, Madonna posited that these components provide such a focal point, the point where waste from a wide spectrum of generators comes together for final processing in a single facility. This is also the point at which large sums of money are collected and accumulated as payment for legitimate treatment, storage, and disposal processes. So, if an organized crime entity would be searching where to have a maximum impact on the movement of hazardous waste, the most logical choice is the treatment, storage and disposal (TSD) facility. An example was presented in the form of a licensed hazardous waste incineration facility that had been forcibly taken over by what was then known as the Genovese Crime Family. After this event, there was never again any genuine attempt to operate a legitimate incineration facility. Accumulated wastes were wantonly dumped and funds taken from waste generators for proper treatment went into the coffers of organized crime. Madonna likened this to a variation of the traditional organized crime scam known as "bankruptcy bailout"; accumulate liabilities (unprocessed waste) and bailout of the company with the assets (cash from processing charges) (Madonna, 1984).

In another case, Madonna described the infiltration of a second hazardous waste facility by a different organized crime family. When the owner of this facility discovered he was dealing with associates of organized crime, he attempted to appease them by offering them an interest in the firm. The organized crime representatives persisted in demanding special favors. This led to the company being drained of assets through traditional methods like "no-show" jobs, special insurance policies, and exorbitant "employee benefits." Organized crime representatives saw to it that the facility's original chemical treatment system was modified through the installation of a "bypass line" that sent millions of gallons of untreated wastes into the city sewer system. Ultimately, this company was rendered insolvent due to the siphoning off of cash receipts by the organized crime associates (Madonna, 1984).

The testimony then moved on to the transportation sequence connecting hazardous waste generators with TSD facilities. In New Jersey, organized crime operatives understood that to achieve a successful takeover of the transporters of hazardous waste in that state, it would be necessary to impose some sort of "coalescing force" to organize and harness hundreds of transporters into a single entity that could be manipulated. This could be represented by a trade association of hazardous waste transporters or an effort to unionize transporters' employees into a "controlled" union. It was pointed out that such techniques had been effectively used by organized crime to take over and impose a system of controls on the *solid* waste industry in the 1970s and 1980s in the northern and central counties of New Jersey. Due to the fact that many of the solid waste collectors had already secured hazardous waste transporters' permits, it was not surprising that an attempt to control the hazardous waste transporters would have its foundation within this group, employing methods already used by the solid waste industry (Madonna, 1984).

To ensure that subcommittee members fully understood the perspective presented in the testimony, Madonna went into a discussion of the system of controls imposed by organized crime on the solid waste haulers known as "property rights." In general, the association was legitimate, adding a charter, a constitution, bylaws, and a legal counsel. In this case, though, Madonna characterized this as simple "window dressing" to hide the imposition and administration of the "property rights" system. The system was described as the right of any association member to have a contract role relationship with the customer for the life of the member. It remains the member's right to sell or to voluntarily relinquish to anyone he chooses. Disputes between members over respective property rights claims would be "adjudicated" by the trustees of the association representing a grievance committee. Those dissatisfied with grievance rulings could turn to their own organized crime "rabbi" as a form of appeal. Preferred members who were more closely connected with organized crime would receive "special treatment." This system was designed to control competition in the industry and allow favored collectors to expand and "dissident collectors" to be controlled and eventually driven from the business. This would often be achieved through unfavorable grievance rulings, threats, sabotage of equipment, and property damage. Ultimately, the property rights system translates into the paying of extremely high rates for garbage collection services and the inability of customers to change garbage collectors. An illustration provided involved a hospital that was "forced" to use a collector that they were dissatisfied with and eventually had to pay twice the price for the "sin" of attempting to secure an alternate collector. The property rights system translates into consumers paying extremely high rates for garbage collection service and being unable to change garbage collectors. Also, if a member collector wishes to sell his business, he has to do so at an artificially inflated price because of protection from competition imposed upon his "stops" (Madonna, 1984).

So how does this fit in with the hazardous waste disposal industry in New Jersey at that time? Madonna visualized, for the subcommittee, a situation in which a trade association/property rights system imposed upon the hazardous waste transporters could serve to seriously undermine legitimate disposal and treatment of hazardous waste. He depicted a situation in which hazardous waste would be transported from and to entities that are priced in line with the rules imposed by organized crime. Gaining control of facilities could seriously divert large volumes of hazardous waste illegally and provide rich profits for organized crime. The primary factor standing in the way of the realization of such a scenario was the relative ease with which individuals contemplating illegal hazardous waste disposal could achieve their ends. Examples presented included how someone could use any type of truck to haul hazardous waste without a performance bond and dispose of the waste at a variety of secluded locations. This, unlike the situation with solid waste in which access to landfills is controlled, the presence of performance bonds, and the purchase of servicing of truck bodies (Madonna, 1984).

A key part of Madonna's testimony is his recommendation of how authorities could stand in the way of any potential takeover of the hazardous waste disposal industry by organized crime as was done in the solid waste industry in New Jersey. After underscoring the difficulties that are inherent in any attempt to de-license a facility or transporter for violation of administrative regulations, he recommended enhanced methods for screening out "undesirables" in the licensing process to begin with. His testimony before the subcommittee allowed Madonna to secure a strong platform for recommending a screening bill, A-901, that eventually became law in New Jersey. The law calls for traditional criminal background investigations for any applicants for hazardous waste facility licenses, hazardous waste transporters' licenses, as well as for solid waste collectors' licenses. The statute dictates that the New Jersey Department of Environmental Protection (DEP) is prohibited from issuing permits for renewing existing permits if investigative reports turn up evidence that applicants possess insufficient reliability, expertise, or competence, or that any person required to be listed on the disclosure statement or identified to have a beneficial interest in the applicant's business: (1) has been convicted of certain crimes within the preceding ten years, (2) does not possess a reputation for good character honesty and integrity, (3) has pending prosecutions or charges pending against him, or (4) has pursued economic gain in an occupational manner or context in violation of the civil or criminal public policy of the state of New Jersey. The law also provides grounds for revocation of hazardous waste licenses previously granted. To address typical property rights strategies of organized crime, the law states grounds for revocation, including coercion of customers and the denial of access to licensed facilities (Madonna, 1984).

Many years have passed since Stephen J. Madonna's testimony was given. Since A-901 has been enacted in New Jersey, there have been attempts to dilute the strength of it. At different points in time, elements in the waste disposal industry have lobbied against the law in an attempt to neutralize certain parts of it. One effort near the end of the year 2011 was somewhat successful. It was alleged by some in the private sector and by some New Jersey public officials that the law was too strict and was scaring off legitimate investors and the jobs they could create. A Virginia company, EnviroSolutions, successfully lobbied for changes in the law making it easier for "secondary" investors, like entities such as hedge funds, private equity firms, and pension funds, to gain stakes in waste disposal ventures by exempting them from fingerprinting and financial disclosure rules. Despite such efforts to weaken this law, at the date of the publication of this book, "A-901" still stands out as one of the country's toughest mob busting laws directed at the waste disposal industry (Pillets, 2011).

References

Cressey, D. (1972). *Other people's money: A study in the social psychology of embezzlement*. Belmont, CA: Wadsworth Publishing Company.

DeVroom, D. (2013, September 24). *Is illegal dumping of hazardous waste a viable business strategy?* Azusa, CA: IDR Environmental Services.

Fagin, D. (2013). *Toms River: A story of science and salvation*. New York: Random House.

Garrison, J., Poston, B., & Christensen, K. (2013, November 16). State fails to keep track of hazardous waste. *The Los Angeles Times*, p.1.

Madonna, S. (1984, February 1). *Organized crime in the hazardous waste disposal industry*. Testimony before the U.S. Senate Committee on Governmental Affairs Senate Permanent Subcommittee on Investigations, Washington, DC.

Martinson, J. (1999, July 21). Royal Caribbean Cruise Line fined $18 million for dumping waste at sea. *The Guardian*. Retrieved from https://www.theguardian.com/world/1999/jul/22/janemartinson

O'Donnell, J. (2013, May 28). Wal-Mart pleads guilty to dumping hazardous waste. *USA Today*. Retrieved from https://www.usatoday.com/story/money/business/2013/05/28/wal-mart-waste/2366999/

Pillets, J. (2011, October 18). Trash firms pushed to weaken laws; oversight eased. *The Record*. Retrieved from http://archive.northjersey.com/news/oversight-eased-on-n-j-trash-industry-1.873262

Rebovich, D. (2015). *Dangerous ground: The world of hazardous waste crime*. New Brunswick, NJ: Transaction Publishers.

Shaikh, N. (2013, April 23). A deadly business [Television Broadcast]. In *Toms River: How a small town fought back against corporate giants*. Cincinnati, Ohio: Taft Entertainment Television.

Taft Entertainment. (1986). *A deadly business*. Thebaut Frey Productions.

Chapter 6

Criminal Methods of Today's Average Environmental Offender

The preceding chapter laid out the dynamics of the driving forces critical to the commission of environmental crime in terms of Donald Cressy's rational choice theory. This chapter delves into the approaches used by environmental offenders to commit those offenses and relies on another criminological theory to explain these approaches. This perspective is provided through the insight of noted criminological scholar Freda Adler in her assessment of methods used by environmental offenders seen through the lens of Cohen and Felson's routine activity theory, which is closely linked to the rational choice perspective (Cohen and Felson, 1979). As Adler explains, Cohen and Felson posit that each criminal act requires the intersection of three elements: *motivated offenders*, *suitable targets*, and the *absence of capable guardians* to prevent the would-be offender from committing a criminal act. Adler envisions an ever-growing body of motivated offenders through the increasing need to dispose of waste generated by industries that produce products like steel, plastics, pharmaceuticals, armaments, and munitions, as well as power-producing enterprises and municipalities. She recognizes that there are numerous targets in the form of airspace (especially surrounding factories), waterways, sewers, isolated dumping grounds, landfills, and abandoned mines (Adler, 1996).

In terms of capable guardians, Adler cites law enforcement officials and the general public who can alert authorities to visible activities of illegal disposal and/or the warning signs of unusual chemical odors. She saves her greatest attention for those enforcement officials responsible for oversight of violations of both regulations and criminal laws and how, in many cases, they have failed to perform in a satisfactory manner. "Among the obstacles to the effectiveness of capable guardianship, there are

> untrained enforcers, difficulty of identifying violators, co-optation of the regulators by those being regulated, financial benefits of paying a fine over complying with standards, ineffective licensing (of ships' captains for example), the imaginative methods of illegal disposal with which enforcement has not caught up, bribery, corruption, weak penalties for violations and too few sanctions". (Adler, 1996, pp. 44–45)

In addition, Adler refers to the tension between regulatory agencies that use cooperative techniques to gain compliance and criminal enforcement agencies, which represent a more punitive form of pressure. She also cites an overreliance by legislators on the manifest system which has proven to be faulty. Topping off her characterization of weaknesses in capable guardians, she notes industries which generate hazardous waste and how they can have a great amount of influence in the shaping of federal legislation involving their own operations (Adler, 1996). By placing environmental crime in the context of the routine activity theory, she ably captures the reasons why certain environmental crime commission methods are favored by offenders and which ones prove to be more perplexing than others to detect, control, and prevent. This chapter follows Adler's lead in using the routine activity theory to frame environmental crime methods as they existed in the past and how they exist today.

General Environmental Violation Categories

Some of the earliest empirical studies of environmental crime methods were conducted, beginning in the 1980s in the United States. The range of methods used to commit these offenses has changed somewhat over time, but the basics of many of these methods remain the same. In the 1980s, those wishing to dispose of polluted substances surreptitiously had a variety of approaches to choose from. Waste generators, haulers, and fraudulent treaters/disposers would often resort to simply disposing of waste through *dispersal on land*. This would be done through dumping in remote areas of vacant lands, wooded areas, or anywhere where the natural environment could supply cover. Sometimes land dispersal would be on land owned by the offender even though they could cause harm to the landowner. Such abandonment of hazardous materials was common in the United States from the 1980s on. Often this abandonment would occur at facilities or businesses that were closing down and simply left for work for the next occupier Abandonment and abandoned mines were prevalent in states like Pennsylvania and West Virginia. In some cases, firms and individuals would commit criminal violations simply by storing hazardous wastes longer than the limit allowed under federal law. For instance, under federal law, it is illegal to store hazardous waste for more than 180 days in the United States without proper treatment/disposal. Hence, some offenders were found to simply "stockpile" these substances and strategically "move" them to stay one step ahead of the law enforcers.

As convenient as land dispersal was in the past and is today, *dispersal into water* was just as popular, if not more. Besides direct disposal into sewer systems, dumping of hazardous substances into waterways such as streams, creeks, tributaries to rivers, bays, and oceans are all attractive means to cunning offenders to illegally dispose. Clearly, *suitable targets* are plentiful when considering bodies of water. Finally, the absence of *capable guardians*, in many cases, enables motivated offenders in their quest to limit expenditures that would be typically devoted to the proper and legal disposal of toxic substances. Such has been the case with government regulators and law enforcers in the present and situations that have developed because of the lack of effective guardianship in the past surfacing in the present. The following material in this chapter will cover the extent to which environmental crime is explained in the context of Adler's framework.

Land Dispersal

Back in the 1980s, New Jersey was a hotbed for the illegal disposal and storage of hazardous waste on land. Wastes from New Jersey as well as from Pennsylvania, Connecticut, New York, Delaware, Rhode Island, and other states would make it to the Garden State as a final resting place for toxic substances generated by manufacturing processes. Chemicals that included cyanide and other flammable, noxious substances would find their way into leased warehouses in places like Jersey City run by firms that would claim to legitimately recycle the wastes and/or treat them where they would actually be stockpiled beyond federal storage duration regulations. Some of the wastes were abandoned under bridges like the Pulaski Skyway in roll-off containers or be crushed at other locations. The system used by the supposed "guardians" was so porous that approved "treatment facilities" sometimes operated for extended periods of time without any on-site incinerators or treatment equipment of any type, simply stockpiling or dumping on their own land. In one instance, wastes were illegally disposed of on facility land with a state environmental inspector on-site. In another case, the state regulatory agency filed suit against city officials because they were seen as acting as "irresponsible landowners" in allowing drums of toxic wastes to be dumped or illegally stored. Some offenders would doctor effluent monitoring devices to give false toxicity readings to state regulators. Other offenders would often "cocktail" hazardous wastes with "waste oil" (i.e., treated used oil that can be reused) and sell as "waste oil" or mix toxic waste with sludge to help mask its harmful content and dump it in landfills. If that didn't work, haulers of tank trucks filled with hazardous waste would be instructed to ride the length of the NJ Turnpike (a little over 120 miles) with their rear spigots slightly open to dispense of the full contents of the tanks over the course of the entire ride. It, indeed, was the "wild west" period of the land waste dumping in the Northeast (Rebovich, 2015). While such land dispersal still exists today in different parts of the United States, its extent and form have been altered somewhat.

One holdover environmental problem from the late 20th century is that of asbestos removal and proper disposal. The Environmental Protection Agency's (EPA's) air toxics regulation for asbestos is aimed at minimizing the release of asbestos fibers during activities involving the handling of asbestos. Clean Air Act (CAA) provisions require the EPA to develop and enforce regulations to protect the public from exposure to asbestos since asbestos contains specific compounds that are known or suspected to cause cancer and other serious health effects like mesothelioma and asbestosis. Asbestos was one of the first hazardous air pollutants regulated under the National Air Toxics Program in 1971. A thorough inspection where demolition or renovation operations occur is required by regulatory law. The regulations require the owner or the operator of the renovation or demolition operators to notify the appropriate delegated entity (often a state agency) before any demolition exceeding the threshold amount of regulated asbestos-containing material. The regulations also require removal of all asbestos-containing materials, wetting asbestos-containing materials, sealing the material in leak-tight containers, and disposing of the waste material as expediently as practical (U.S. EPA, Asbestos NESHAP, 2016). Since the 1970s, individuals have navigated around these regulations, either motivated by a need to save money on the requisite costs of treating/disposing of these wastes properly or to make money by contracting out services to "hide" the wastes at "suitable targets." These circumstances still manage to crop up in the 21st century.

One of these situations took place in upstate New York on a 28-acre piece of property near the Mohawk River in Herkimer County, N.Y. The isolation of this area made it desirable for the clandestine dumping of the waste. When the mounds of construction debris were found by investigators, they also found evidence that children had routinely ridden their bikes past them. Much like cases found in Maryland as part of the Northeast Hazardous Waste Research Project in the 1980s and 1990s, this dumping involved a conspiracy between those wishing to illegally dispose of hazardous wastes and the owner of farmlands willing to supply a "suitable" destination on the part of his own property. In this particular case, it amounted to a conspiracy between operators of a solid waste recycling and management facility located in central New Jersey and a property owner over 250 miles away in upstate New York. The conspiracy resulted in the illegal dumping of asbestos-contaminated construction debris that totaled thousands of tons of material. The waste was comprised of demolished buildings that had been shredded but without extracting harmful asbestos. The scheme seemed so foolproof to the offenders that long-term plans were made to continue the criminal operation for up to a five-year period. Enabling the offenders to conceal their criminal activity was their ability to take advantage of the state regulatory agency through the fabrication of a state permit and forging the name of a regulatory official on the fake permit. The offenders' operations finally began to unravel when a trucker who had hauled some of the waste became suspicious that the site wasn't authorized to accept such waste and reported that to authorities. After the offenders caught on that a subsequent investigation had begun, they scrambled to conceal critical documents and submit falsified documents of environmental sampling to the EPA. But it was too late. Sentences were meted out to offenders amounting to 36 months in prison each and restitution of close to $500,000 (USDOJ, August 2, 2013 . The state of New York proved to be the site of other asbestos-related land disposals between 2011 and 2012. In this case, the owner of a New York real estate holding company had ordered workers to illegally remove thermal system installation that contained asbestos from buildings owned by him and occupied by different tenets. This individual conspired with a subordinate manager to instruct a maintenance employee to remove this material from piping without properly informing the individual of the harmful nature of the material or providing him with protective equipment. The material was transported in open bags in the uncovered bed of a truck and illegally stored in a U-Haul style box truck in a shed maintained by the city's office of public works (Intellihub, 2014).

As stressed by Adler, the lack of effective guardianship can do much to enable the criminal actions of those bent on committing environmental crimes. But what of those working to serve the protection of the public and the environment actually conspiring with offenders? Such was the case in 2014 in the Buffalo area of New York State, where a New York State certified air sampling technician and a project monitor were convicted and sentenced in a case involving an asbestos abatement project in Buffalo's Kensington Towers. For six months, contractors conducted asbestos abatement activities at six buildings in Buffalo's Kensington Towers complex. The two technicians were certified by the New York State Department of Health to conduct asbestos project monitors and air sampling duties. Knowing that those conducting the asbestos abatement work were violating federal regulations, the government contractors aided and abetted violations by conducting visual inspections and final clearance of air sampling that indicated no violations of the asbestos work practice standards. In doing so, they acted as accessories to the false statements of the other defendants (Figure 6.1) (U.S. FBI, 2014).

It would be extremely myopic to conclude that asbestos-related environmental offenses have been confined to only New York State. In 2015, five individuals were convicted and received the statutory maximum of 5 years in prison for the illegal removal and disposal of asbestos-containing materials at what was the former Liberty Fibers plant in Hamblen County, Tennessee. Once again, this case involved bankruptcy in that the salvage company contracted to do the removal by purchasing a bankrupt facility in the effort to salvage metals that remained in the plant after its closure. The irony here is that during the sentencing hearing, medical experts testified that it was a

Figure 6.1 Kensington Towers before demolition. www.epa.gov/enforcement/2015-major-criminal-cases

substantial likelihood that the salvage workers could suffer death or serious bodily injury as a result of their exposure (U.S. DOJ, January 23, 2015).

As our society continues to grapple with past mistakes made through the inclusion of asbestos in buildings, we are still confronting the lingering issue of the removal of asbestos from existing structures. Be it hospitals, courthouses, schools, or other public buildings, we will continue to experience the need for the professional and legal removal of asbestos. Along with this is the need to dispose of these wastes from asbestos removal in an effective and legal manner. The end is not yet here for this chapter of asbestos removal, but we are getting closer. As long as there is still a need for asbestos removal, there will be enterprising individuals who will seek to take advantage of weak guardianship systems and dispose of asbestos wastes in the most criminally expeditious manner.

In the southern region of New York State, other offenders have been known to use their property for the illegal handling of other types of hazardous wastes. In 2013, the owner of a firm in Owego, New York, that was in the business of removing old industrial plating equipment for reuse or recycling entered into an agreement with a bankrupt waste generating facility in New Hampshire to remove hazardous chemicals that included arsenic, chromium, lead, and selenium. This particular profit-motivated offender was not properly licensed to handle such wastes but did so anyway, skirting the law and law enforcement officials. Over the course of a year, this offender illegally stored hazardous waste without proper labeling and did not isolate incompatible materials, raising the possibility of combustion of interacting chemicals. His interpretation of "treatment" was to recklessly ignite wastes and mix other wastes together and ship to off-site locations. This was another case of offenders exploiting monitoring procedures by not appropriately using manifests as required (WBNG News May 13, 2015).

Figure 6.2 Workers clean floors in the West Calumet Housing Complex in East Chicago, Indiana. www.epa.gov/uss-lead-superfund-site/uss-lead-photo-gallery

And when considering past societal mistakes, one must not lose track of the long-term effects of callous actions of land dispersal when such actions were once considered benign. An area just north of a huge former USS Lead smelting plant was at the center of a controversy in August of 2016. At that time, the mayor of East Chicago, Indiana, told over 1,000 poor minority residents that they had to move out to do high lead content in the soil. In 2008 and 2011, the EPA removed soil from the hot-spot areas. The EPA sued the companies responsible for the contamination in 2009, and by 2012 had a cleanup plan to remove all contaminated soil. However, testing to determine which soil needed to be removed did not begin until November 2014. Further, the EPA did not receive the final results showing the exact location of the contamination until May 2016. This was reportedly a result of a disagreement with the contractor the EPA hired to tabulate the data and concerns about the validity of the data. While this is not an example of environmental criminality per se, it does underscore the failures of deficiencies in the *guardianship* factor and does deserve to be mentioned in the context of Adler's paradigm protection from environmental threats. In the final analysis, such land contamination becomes the responsibility of protection through our society's guardians, and, if inefficient, can result in the undue suffering of innocent citizens. With land dispersal of hazardous substances, the sins of our past continue to haunt our present and promise to continue to do so in our future (Figure 6.2) (Goodnough, 2016).

Mining-Generated Land and Water Dispersal

Dispersal of hazardous materials through mining can be separated into subcategories of mining abandonment and active mining. Another subcategory which is not considered traditional mining is drilling into the earth to access natural gas in the process commonly known as fracking.

One of the largest abandoned mine environmental cases involved the Chevron Questa Mine that operated from 1919 through 2014, at which time the mine was permanently closed. Throughout that period, mining operations and waste disposal were responsible for contaminating surface water and groundwater. During the time the mine was in full operation, over 325 million times of acid-generating waste rock were excavated and disposed of in several enormous waste rock piles. The mine was located in the state of New Mexico. In this case, the EPA reached a settlement with Chevron in which Chevron is required to pay over $140 million to clean up the site covering 275 acres of a tailing facility. At this facility, mine wastes were stored in a water treatment plant, and a groundwater extraction system was built. Ultimately, through settlement, Chevron reimbursed EPA for over $5 million to clean up the site. In addition to cleanup activities at the site, Chevron Technology Ventures constructed a concentrating photovoltaic (CPV) facility on a 20-acre portion of the site. The CPV facility's 173 solar panels automatically track the path of the sun and, over the course of a year, produce about one megawatt of energy. This initially gave the appearance that it would represent one of the more amicable conclusions to a mining pollution case (Figure 6.3). Other cases involving mining wastes did not go so harmoniously (DOJ, August 9, 2016).

Another similar case involving abandoned mines was more far-reaching and dealt with a substance that was considered extremely harmful: uranium. This case involved over 27,000 square miles within Utah, New Mexico, and Arizona in the Navajo Nation territory. This region is rich in uranium, which is a radioactive ore that became in high demand with the development of atomic power and weapons at the end of World War II. Between 1944 and 1986, over 4 million tons of uranium ore were extracted within the Navajo Nation. Countless Navajo families lived in proximity to these mines. In a 2015 settlement agreement with the Navajo Nation, the U.S. EPA reached an agreement to provide cleanup of more than 30 abandoned uranium mines at a cost of over $100 million. However, by August 2016, leaders of one of the nation's largest American Indian tribes blasted the EPA as the tribe's attorneys sued the EPA claiming negligence in the

Figure 6.3 The Chevron CPV facility. www.epa.gov/superfund-redevelopment-initiative/superfund-sites-reuse-new-mexico

cleanup of a massive mine waste spill that tainted rivers in the three western states. While this is not considered a criminal case, it calls to attention the serious problem that abandoned mines pose to the general public in the United States (Figure 6.4) (U.S. DOJ, August 9, 2016).

One of the most egregious cases of abandoned mine pollution took place in 2015 and accentuated the failure of effective guardianship to control such offenses. On August 5, 2015, an EPA crew at the Gold King Mine in southwest Colorado accidentally released over three million gallons of water tainted with high levels of mercury and arsenic. This toxic water emanating from an abandoned gold mine emptied into the Animas River, turning it a yellowish-orange color. The Animas River is an important river since it is part of the Colorado River system supplying over 20 million people with water and irrigating over three million acres of farmland. The waste subsequently flowed into New Mexico's San Juan River and then into Lake Powell, costing the EPA close to $30 million to correct. A year later, in August of 2016, the EPA was still found to be wrestling with controlling the wastes. Originally, the EPA was in Colorado to halt the toxic flow from the abandoned mine, which had leaked at a rate of 50 to 250 gallons a minute for years. The EPA's excavation turned out to be a monumental failure, resulting in the increase of wastewater release at a rate of 740 gallons a minute. The EPA built two treatment pools to filter the flowing wastes; however, it was discovered that there were several other local abandoned mines that had been discharging similar wastewater for decades. These releases had never been filtered through

Figure 6.4 Sampling at one of the abandoned uranium mines in the Navajo Nation territory. www.epa.gov/navajo-nation-uranium-cleanup/cleaning-abandoned-uranium-mines

a treatment plant. This also happened to have been another case in which the local government had played the "rationalization game." Knowing of these discharges, the local government had avoided applying for Superfund status for the area, fearing a drop in tourism and property values. As a result of the Gold King Mine release and the bungling efforts of the EPA exacerbating the problem, the local government was finally compelled to apply for Superfund status, some experts believe, too little too late. As pointed out by a representative from the environmentally conscious Denver-based Keystone Policy Center, the Gold King Mine is but one of tens of thousands of abandoned mines in the western United States continually polluting the environment (Figure 6.5) (Hood, August 4, 2016; *The Economist*, August 15, 2015).

Hydraulic fracturing or fracking is a means of natural gas extraction employed in deep natural gas well drilling. Once a well is drilled, millions of gallons of water, sand, and proprietary chemicals are injected, under high pressure, into a well. There are a number of these active deep injection wells for oil and gas waste in Pennsylvania. Typically, the oil and gas industry uses injection wells to dispose of wastewater. The wastewater contains chemicals, heavy metals, and radioactive material. Much of the frack water produced in Pennsylvania gets trucked to Ohio, which has more disposal wells. In 2014, a criminal matter developed regarding fracking wastewater that is emblematic of the types of offenses we may see more of in the future if effective monitoring is not in place.

In March of 2014, a former employee of a Youngstown, Ohio, oil- and gas-drilling company was sentenced to 3 years' probation for dumping tens of thousands of gallons of fracking waste into a tributary of the Mahoning River. The Mahoning River is a river located in eastern Ohio and western Pennsylvania and joins the Shenango River to form the Beaver River. The Mahoning

Figure 6.5 Cleanup response to Gold King Mine incident. Treatment ponds were built at mines close to the Gold King Mine. When water would leave the mine, these ponds would slow it down and allow adjustment of the pH and let contaminants settle to the bottom. www.epa.gov/goldkingmine

River is also part of the Ohio River watershed. At his hearing, the polluter contended that he was following orders from his supervisor for fear he would lose his job if he failed to comply. He ultimately lost his job after the excavating company went out of business. The owner was sentenced to 28 months in prison for ordering the dumping. The owner had ordered employees to run hoses from 20,000-gallon storage tanks to a nearby stormwater drain and open the release valve, sending waste liquid into the Mahoning River. In this case, the employee testified that the owner ordered him to perform the secret dumping under cover of darkness and after all of the other employees had vacated the facility. The employee added that the owner instructed him to lie if questioned about the dumping and to tell law enforcement officers he had emptied the waste tanks only six times when it, in fact, had been emptied more than 30 times. The *guardian* in this case was an alert citizen who reported the dumpings to authorities. Court records reflected that the prosecutor stated the tributary creek into which the wastes were dumped was "void of life" and that "Even the most pollution-tolerant organisms, such as caddis flies, were not present. The creek was essentially dead" (Caniglia, 2014). This type of offense mirrors environmental crimes from decades past in which crooked waste handling employers identify employees they believe more inclined to comply with unlawful directives, isolate them from other employees, and try to ensure that they commit the offenses out of public sight. Historically, this method was a mainstay scenario of corrupt treatment and disposal facility operators from the past. The main aspect that has changed is the source of hazardous waste: fracking.

It is expected that fracking will remain a potent issue with environmental groups in the United States for the future. In 2015, the United States had 300,000 fracking wells, far more than the 23,000 in 2000. Besides the risk of pollution from mishandled fracking wastes, there is some speculation that the concentration of such operations in certain parts of the United States has raised the chances of resulting earthquakes. Some have pointed to the 5.6 magnitude tremors in September 2016 in central Oklahoma as being a result of fracking operations. Months prior to the earthquake, government scientists warned that oil and natural gas drilling had made a wide swath of the country more susceptible to earthquakes. A March 2016 report by the U.S. Geological Survey on "induced earthquakes" said as many as 7.9 million people in parts of Kansas, Colorado, New Mexico, Texas, Oklahoma, and Arkansas were facing the same earthquake risks as those in California (Wattles and Egan, 2016).

Water Dispersal: Ocean-Going Vessels

The Environmental Crimes Section of the U.S. Attorney's Office reported that one of the greatest recent threats to the environment has been the threat posed by ocean-going vessels. The operation of these vessels generates large quantities of waste oil and oil-contaminated wastewater. International and U.S. law requires that these vessels make use of pollution prevention equipment to preclude the discharge of such wastes. If any overboard discharges occur, documentation is mandated in an oil record book, a law that is regularly inspected by the U.S. Coast Guard. In 2014, the monetary penalties that were imposed in pollution cases involving ocean-going vessels totaled over $11 million. Since the late 1990s, these types of cases accounted for more than $352 million in criminal fines and more than 27 years of imposed imprisonment. Some of the most deliberate of these cases included Noble Drilling LLC, M/T Bow Lind, M/V Trident Navigator, M/V Thetis, and M/V Neameh.

In the Noble Drilling case, the drillship Noble Discoverer and the drilling unit Kulluck conducted illegal drilling operations off the coast of Alaska. In 2012, Noble experienced problems in the management of bilge and wastewater that had collected in the engine room spaces of the Discoverer. Noble addressed this problem by constructing a makeshift barrel and pump system and illegally discharging wastewater into the ocean. In addition to this, the company pumped hazardous skimmer tank fluids into ballast tanks and then illegally discharged them overboard. Those individuals responsible for this also failed to record the transfer to the ballast tanks and the subsequent discharges in the record book as required by federal regulations to, in essence, help cover up the violations.(U.S. DOJ, 2016).

The Noble case became a prime example of the operations of an ocean-going vessel that created and maintained a culture of environmental criminality that spanned many different areas. Exacerbating the actions described above, Noble failed to notify the Coast Guard of a number of serious hazardous conditions that were occurring aboard the Discoverer. The ship experienced many problems with its main propulsion system, including its main engine and its propeller shaft. This resulted in a number of engine shutdowns and equipment failures, leading to unsafe conditions aboard the vessel. In some instances, problems experienced with the ship's main engine resulted in high levels of exhaust in the engine room. Such problems also resulted in fuel and oil leaks and backfires. The Kulluck, also owned by Noble, eventually ran aground off the coast of Unalaska during a bout of bad weather. Due to the failure of its main engine and other equipment, the Discoverer had to be towed from Dutch Harbor to Seward because of the many failures of its main engine. As a result of all these violations, Noble Drilling was required to pay $12.2 million in fines and community service payments and, additionally, the survey for their probation. During the probation time, the company was required to implement an effective environmental compliance plan (Figure 6.6) (U.S. DOJ, 2016).

An earlier ocean-going vessel pollution case was discovered during a November 2012 Coast Guard inspection of the M/T Bow Lind, a petroleum/chemical tanker ship. The inspection

Figure 6.6 Noble Drilling Unit Kulluck. www.epa.gov/enforcement/2015-major-criminal-cases

revealed that the ship had been discharging oily bilge water into international waters. It was found that this action was routinely done multiple times. Evidence pointed to the chief engineer of the ship directing crewmembers to purposely bypass pollution prevention equipment. Concealment of these actions was achieved through deceptive entries in the vessel's oil record book, along with complete omissions of actions. The chief engineer was sentenced to serve three months' incarceration, and the company was fined over $900,000 (U.S. DOJ, 2016).

Similar actions were conducted on another ocean-going vessel, the M/V Trident Navigator, approximately one year later. The chief engineer of that vessel directed crew members to create a bypass to illegally dispose of oily bilge waste. Circumventing the ship's oil water separator and oil content monitor, several metric tons of waste were discharged. As with the M/T Bow Lind case, the illegal discharges were never recorded in the oil record book. Further, the chief engineer was found to have confiscated a crewmember's cell phone and deleted a photograph of the illegal bypass. In January of 2014, Coast Guard personnel came aboard the ship while it was anchored near New Orleans. The Coast Guard took this action after they were tipped off by a crew member. The criminality of the chief engineer was found to be even more serious as investigators dug into the case. They found that the chief engineer had instructed crewmembers to deny any knowledge of the bypass in an effort to obstruct the investigation. Ultimately, the company was fined over $800,000 (U.S. DOJ, 2016).

A similar example of ocean-going vessel water dispersal of hazardous wastes was evidenced in another case in which managers aboard the vessel engaged in actions involving the direction of subordinates to commit criminal acts and the attempts to conceal these actions. In the case of the M/V Thetis, charges of criminal conspiracy were also leveled at leaders on the ship. From 2011 through 2012, the chief engineer and the second engineer routinely ordered illegal discharges of sludge and bilge wastes and then directed others to hide bypass piping and to falsify the ship's oil record book. The two were eventually convicted of criminal conspiracy at trial, obstruction of justice, and falsification of records. Both were sentenced to one-year terms of probation, and the company was fined over $1 million (U.S. DOJ, 2016).

A relatively large portion of cases involving ocean-going vessels illegally disposing in United States waters or off the coast of the United States have been committed through vessel operators from other countries. The most recent case, ending in September of 2016, resulted in the conviction of Oceanic Illsabe Limited (Oceanic Illsabe), Oceanfleet Shipping Limited (Oceanfleet Shipping), and two of their employees with violating the Act to Prevent Pollution from Ships (APPS), obstruction of justice, false statements, witness tampering, and conspiracy in relation to oily waste dumping from the M/V Ocean Hope into the Pacific Ocean. Both of these operations emanate from Greece. Like the M/V Thetis case, a conspiracy between the chief engineer and the second engineer was at the heart of illegal discharges into water. The second engineer was guilty of bypassing pollution prevention equipment with an unauthorized hose connection, or "magic pipe" used to discharge oil sludge into the sea. In numerous instances, crewmembers were directed to pump oily mixtures from the vessel's bilges through the use of the vessel's general service pump rather than properly processing the substances to the vessel's pollution prevention equipment. A fictitious oil record book was maintained to hide the illegal discharges. Crewmembers were also advised by the chief engineer and the second engineer to conceal illegal activities through lying to representatives of the U.S. Coast Guard. At the time of the printing of this book, sentences had not been imposed, but the companies were subject to being fined up to $500,000 per count, in addition to other possible penalties imposed. The chief engineer and the second engineer faced a maximum penalty of 20 years in prison for obstruction of justice (U.S. DOJ, December 11, 2015).

This case was not the first time the use of a "magic pipe" or "magic device" had been used in a criminal case involving foreign ocean-going vessels. In 2016, a Filipino citizen and captain of the tanker ship T/V Green Sky pleaded guilty to one felony count in federal court in Charleston, South Carolina, resulting from a U.S. Coast Guard investigation into pollution crimes aboard the vessel. The person pleading guilty was the highest-ranking officer aboard the ship and admitted that members of the ship's engine room, including a senior officer, illegally discharged polluted substances overboard. A bypass of the vessel's oily water separator was not recorded in the oil record book, making it a federal crime upon entry into a U.S. port. The vessel was registered in Liberia and is owned by an entity incorporated in the Marshall Islands (U.S. DOJ, February 18, 2016).

In another 2016 case, a Norwegian shipping company, DSD shipping, was fined $2.5 million for illegally disposing contaminated oil into the ocean off of Mobile, Alabama. The criminality in this case ran even deeper with convictions for jury tampering and revelations that previous deficiencies in the operation of pollution prevention equipment had been identified in other DSD vessels while they were in international ports. Like the other cases, falsified entries in the vessel's logbooks were discovered. Such was the case with Ciner Gemi Acente Isletni Sanayi Ve Ticaret S.A., a ship management company based out of Turkey. Ciner operated M/V Artvin, a 44,000-ton bulk carrier that transported cargo from ports around the globe. Discharges were detected in the Port of Baltimore, and it was found that the firm routinely dumped from the vessel into the sea without the use of required pollution prevention equipment. The chief engineer and the second engineer, both from the Philippines, directed crewmembers to dump oily water from waste oil tanks into the sea without using the vessel's oil-water separator. Once again, false entries were made into the vessel's oil record book. Through a plea agreement, the company was fined over $1 million (U.S. DOJ, April 8, 2016).

Offshore Oil Platforms

An area ripe for environmental pollution centers is the operation of offshore oil platforms. The oil and natural gas beneath the United States' ocean floor represents an energy windfall to some and an environmental threat to others. Since the BP/Deepwater Horizon disaster, the number of platforms has declined, but there are still over 30 rigs in the Gulf of Mexico. The two basic problems from offshore drilling, from the perspective of environmentalists, are pollution from everyday operations and oil spills from platforms, pipelines, and tankers. We are now seeing cases in which operators act unethically to save spending on proper treatment/disposal of wastes. In a case settled in 2015, the U.S. Department of the Interior's Bureau of Safety and Environmental Enforcement (BSEE) and the U.S. EPA and ATP Oil & Gas Corp agreed to resolve actions under the Clean Water Act (CWA) and the Outer Continental Shelf Lands Act (OCSLA) concerning unauthorized discharges of oil and chemicals from a floating oil and gas production platform into the Gulf of Mexico off the coast of Louisiana. The agreement led to the imposition total of $41.85 million in judicial and administrative penalties for the violations. ATP had been accused of discharging oil and an unauthorized chemical dispersant into the Gulf of Mexico from ATP's oil and gas production platform known as the ATP Innovator. A regulatory inspection of the ATP Innovator

Criminal Methods of Environmental Offender   73

in March 2012 uncovered alleged unlawful discharges of an oil and a piping configuration that routed an unpermitted dispersant – a chemical mixture to break up oil – into the facility's wastewater discharge pipe to mask excess oil being discharged into the ocean. At the time, ATP was the operator of the facility, and ATP Infrastructure Partners (ATP-IP) was the non-operating owner. The ATP Innovator was located in the Mississippi Canyon area off the Gulf of Mexico, 45 nautical miles off of southeastern Louisiana. In 2013, the platform was removed from the deepwater production site and towed to port in Corpus Christi, Texas (U.S. DOJ, November 19, 2015).

Air Dispersal

While up to now the primary focus of this chapter has been on violations involving land and water dispersal, one should not lose track of the area of the improper dispersal of pollutants into our air. A primary example of this type of case ended in 2015 with the imposition of a $2.25 million federal civil penalty against the General Electric Company. This case centered on GE's use of an incinerator at a manufacturing facility that was once owned and operated in Waterford, New York. Between 1947 and 2007, GE manufactured a wide variety of products at the facility. These products included sealants made of silicone, which generated hazardous waste. GE received permits from New York's Department of Environmental Conservation (DEC) to dispose of hazardous wastes on-site. The company was to dispose of hazardous wastes in a rotary kiln incinerator. This incinerator had an automatic waste feed cut that was designed to shut down if the company deviated from operating parameters designed to ensure compliance with the Clean Air Act and the Resource Conservation and Recovery Act. To allow GE to burn hazardous waste in the incinerator in violation of these two acts, a computer program was used to override the incinerator's automatic waste feed cut-off system. It was found that in over 1,800 instances, employees were instructed to override the system in 2006 and 2007. Consequently, the public was exposed to harmful air pollutants, including dioxins and furans. To cover up these practices, GE managers routinely submitted false compliance reports to the United States in the state of New York (U.S. DOJ, December 3, 2015).

Discussion

The main objective of this chapter was to describe some of the most common methods used by those who would break environmental laws by polluting ground, water, and air. As a foundation for analysis and discussion, the chapter relies upon Adler's triangle of three elements necessary for environmental crimes: *motivated offenders, suitable targets*, and the *absence of capable guardians*. The book *Dangerous Ground* characterized the heyday of environmental crime in the United States in the 1980s, where these three elements came together to facilitate the proliferation of environmental crime, particularly in the Northeast. Many of the cases examined in that book centered on crimes committed by small businesses, public institutions, and enterprising treatment/storage/disposal (TSD) facilities. For small businesses and public institutions, a majority of the offenses were committed by situational offenders, those individuals who took advantage of a particular situation to cut costs or totally eliminate costs of properly treating hazardous wastes. In the case of TSD facilities, their crimes tended to be more entrepreneurial offenses in that a large portion of these criminal operations were engaged in the improper disposal of hazardous wastes as a way to *make* money by accepting wastes to treat at a high cost to waste generators and simply finding ways to surreptitiously avoid their own expenses of properly treating the wastes by disposing of the wastes untreated. In a big way, this became a part of "doing business" for them (Rebovich, 2015).

As we examine more recent environmental crime in the 21st century, we find that the influences of the three elements of Adler's environmental crime triangle have not changed drastically, but the sources of criminality have changed somewhat. For many years, buildings in the United States have contained asbestos, which was used in spray-applied flame retardant, thermal system insulation, and in a variety of other materials. Structural components like asbestos panels were also used, and asbestos was sometimes "flocked" above false ceilings, inside technical ducts, and in many other small spaces where firefighters would have difficulty gaining access. Since fully discovering the harmful effects of breathing filaments of asbestos, we, as a society, have been committed to asbestos removal in buildings. The glaring problem exists in the form of what happens to the asbestos once it is removed and once buildings containing asbestos are demolished. Legally, the asbestos wastes must be properly treated and disposed of. This chapter has stressed that, depending on the situation, those responsible for handling asbestos have found ways to avoid paying the high cost of properly treating and disposing of this hazardous waste. Some cases involved interstate conspiracies between those possessing the waste and those willing to dispose of the wastes on their own private property where the expansiveness of the property (e.g., farmlands) could serve as a prime *suitable target* for clandestine dumping. Incidents like these were successfully engineered with cover provided by the age-old method of the fabrication and forgery of documentation that would verify the proper handling of wastes, in effect neutralizing the efforts of *capable guardians*. One asbestos-related case highlighted how the co-opting of *capable guardians* plays a part in the commission of asbestos disposal crimes. In that case, regulators lied to cover violations of offenders and an asbestos abatement project and were accessories to false statements. In his discussion of environmental criminality, the famed criminological scholar Gerhard Mueller noted how frequently a symbiosis can develop between regulators/enforcers and those regulated, which, for the sake of smooth working relationships, sacrifices the common good (Mueller, 1996).

Bankruptcies were found to be the "situation" for other situational crimes against the environment. Some of these cases were asbestos related, while others involved hazardous chemicals that were generated through industrial manufacturing operations. In both general categories, these were cases of abandonment in which buildings were simply left unattended for others to worry about. As explained in *Dangerous Ground*, this represents the ongoing "game" of environmental "hot potato" in which the ultimate "loser" is the very last entity in possession of the disposed wastes (Rebovich, 2015). Sometimes those "others" happened to be low-income citizens, populating areas surrounding the abandoned operations like in the Indiana city of East Chicago. In one particular case, the failures of effective guardianship led to inexcusable delays and cleaning up contaminated areas and, thus, resulted in extended exposure to threats to personal health.

A different type of environmental abandonment has been found to plague the mining industry in the 21st century. One particular case involved wastes that had been generated through close to a century of mining operations. Effective guardianship was found to be wanting in two cases of mining abandonment. The first case centered on uranium mining operations on Native American territory in which the U.S. EPA was sued for negligence. The second example proved to be an embarrassing "black eye" for EPA guardianship in that EPA efforts to effectively address the abandoned mine pollution actually exacerbated the problem, creating a much more serious environmental problem than what was originally detected. Both cases serve to underscore the importance of what Adler calls *effective guardianship* needed to contain and prevent environmental pollution. Adding to problems like these are situational efforts by local governments to deny the seriousness of toxic dumping/spills regardless of whether they are intentional or not. Often these efforts are designed to rationalize away environmental threats to preserve businesses and property values of areas that have been affected. In that sense, these actions play into the problem of ineffective guardianship.

Actions to take advantage of advances in tapping into newer sources of oil and gas deposits have also led to increased pollution. While offshore platform drilling for oil has been declining recently, existing wells present threats to the environment, especially when steps are taken to dispose of wastes illegally into waterways. Hydraulic fracturing, or "fracking," has led to instances in which fracking operators ordered subordinates to dispose of wastes illegally. The methods used here harken back to previous decades' practices at TSD facilities in which criminal dumping assignments are given to criminal "core group" members to conduct "midnight dumpings" under cover of darkness to avoid detection. While the goals of the legitimate operations in fracking are quite different from the goals of past TSD facilities, it appears that the methods for illegal dumping have largely stayed constant.

The chapter points out that a growing concern for those wishing to protect the environment revolves around criminal activities on ocean-going vessels. The case studies presented in the chapter highlight the extent to which the offenders will go to commit the offenses and then cover them up. Of special interest is how managers on the vessels would direct subordinates to illegally dispose of waste through circumventing or otherwise bypassing mechanisms in place to properly treat the wastes before they were disposed. These actions were very similar to actions taken at TSD facilities in the 1980s in which facility yard workers were ordered to "doctor" treatment devices such that the appearance was given that wastes were being properly treated when, in fact, no such treatment was being conducted at all. Like those earlier cases, ocean-going vessel cases included not just the conspiracies between managers (in these cases, chief engineers and second engineers) but also the blatant falsification of records and direct instructions to subordinates to lie to government investigators particularly with regard to the existence of illegal bypassed systems.

In the final analysis, it is clear that, as a society, we have made advances in protecting our environment from environmental offenders. Gone are the days when operators of TSD facilities ran roughshod over environmental laws and environmental law enforcement in a widespread pattern of criminally disposing of untreated wastes. However, the synergy between *motivated offenders* and *suitable targets*, in the *absence of capable guardians*, continues to this day. The suitable targets are relatively unchanged. The motivation for offenders is less entrepreneurial than had been in the past but exists nonetheless in cases in which individuals feel they are in situations where they must cut operating and treatment costs through simple means of criminal acts. And, in too many cases, our *guardians* enable these criminal activities to take place through complacency, co-optation, and incompetence.

Case Study: Special Problem in Water Dispersal – "Ship Breaking"

This chapter pointed out that one of the most critical present threats to the environment has been the threat posed by ocean-going vessels. The operation of these vessels generates large quantities of waste oil and oil-contaminated wastewater that are often wantonly dumped into the ocean. But dig a little deeper and one finds a subcomponent of this topic that is alarming, to say the least. It is something called "ship breaking." Ship breaking is really another name for ship demolition. This type of demolition entails the dismantling of ships for either one of two reasons. First, it can be done to salvage ship parts that can be sold for reuse. Second, ship breaking can be conducted to extract raw materials, primarily "scrap." Typically, the lifespan of ships is between 25 to 30 years before they succumb to metal corrosion. A majority of these deteriorating vessels are ultimately run ashore in developing countries for disassembly. When this occurs, the waters are subjected to the disposal of asbestos, lead, and polychlorinated biphenyls (PCBs), among other pollutants (NGO Shipbreaking Platform, February 3, 2014).

The NGO Shipbreaking Platform, a global coalition of organizations dedicated to the prevention of dangerous ship-breaking practices around the world, has reported that of the 1,213 large ocean-going vessels that had been scrapped in the year 2013, 645 were sold to substandard beaching facilities in the countries of Bangladesh, Pakistan, and India. It was estimated that close to 40% of these ships were owned by European Union (EU) entities. The preferred dumping ground for these deteriorating ships has been South Asia, where there is poor enforcement of environmental safety and labor rights standards. In 2013, a total of 372 large commercial vessels owned by European ship owners were sold for the purpose of breaking them up. Of that number, 238, almost two thirds, wound up on beaches in South Asia. Of the offending ship owners, the worst offenders appeared to emanate from Greece. Second in line for this dubious distinction was Germany. A record 80% of these ships ended up in India, Bangladesh, and Pakistan. Some of the offending companies were well-known – Danaos and Euroseas (Greece); Conti, Hapag-Lloyd, and Leonhardt & Blumberg (Germany). Other offending dumpers of this kind included the Mediterranean Shipping Company (MSC) out of Switzerland (a firm that dumped nine ships in India) and the Monaco-based Sammy Ofer Group (a firm that disposed of 13 ships in Bangladesh, Pakistan, and India) (NGO Shipbreaking Platform, February 3, 2014).

Ships broken in Bangladesh, Pakistan, and India are of particular concern to the environment. This is because the broken ships are dumped on beaches with sand so soft that they cannot support safety measures like heavy lifting or emergency response equipment. Ship-breaking pollutants seep directly into the coastal zone environment. Such ship breaking is not permitted on the beaches of more developed countries. It is important to understand that it is feasible for ship breaking to be completed in a safe manner as long as proper technologies are used and regulations strictly enforced. Unfortunately, polluting alternatives are used quite often for cost savings. On the beaches of South Asia, ship-breaking practices expose those who conduct the ship breaking to harmful pollutants. Many of the workers are migrant workers and children who conduct the ship breaking activities without masks to protect their lungs from asbestos and poisonous fumes and, in some cases, without shoes. The International Labor Organization (ILO) has stated that ship breaking occurring on beaches represents one of the most dangerous jobs to workers. In addition, it is pointed out that the ships that are sold to ship recycling facilities often are not dismantled in a manner in which workers are substantially protected. In many cases, these facilities do not possess satisfactory technology, nor do they stress enhanced expertise and training. The traceability of these wastes at the facilities remains quite vague (NGO Shipping Platform, February 3, 2014).

In November of 2013, The European Parliament and the Council of the European Union adopted the Ship Recycling Regulation to respond to the growing problem of ship-breaking pollution. The primary goal of the regulation was to reduce the negative impacts linked to the recycling of ships from the EU. This regulation "piggybacks" onto the requirements of the 2009 Hong Kong Convention for the Safe and Environmentally Sound Recycling of Ships. This regulation also includes additional safety and environmental requirements authorized by Article 1(2) of the Convention. The specific objective of this regulation on ship recycling was to take aim at practices like ramming giant tankers full of toxics into beaches. The problem was that registering European ships under "flags of convenience" was far too easy and provided a loophole in the legislation for the biggest and most toxic tankers. Under the 2013 legislation, the breaking of ships registered under the flag of an EU member state in beaching yards without proper recycling processes in place was prohibited. The regulation was seen by many as a "paper tiger." It was found that over 65% of the European ships dismantled in 2013 refrained from sailing under the flag of an EU member state when bound for a dismantling destination and was, thus, not covered by the new regulation. Over 50 ships were "flagged out" from European registries prior to being scrapped outside the European Union. The most popular flags flown for "end-of-life" ships that were ultimately "broken" on beaches were from the islands of Comoros, Tuvalu, Saint Kitts and Nevis, and the West African country of Sierra Leone (European Commission, 2016).

As a result of the introduction of the Ship Recycling Regulation, certain responsible European ship owners have practiced sound ship recycling policies. An example is the Maersk Group from Denmark. However, that same shipping company eventually sold three of its ships to the Greek company Diana Shipping and chartered the ships back. Ultimately, these ships were unscrupulously beached. This practice is not uncommon. One way of avoiding responsibility for end-of-life vessels is to sell off to a new owner while continuing to be the operator. Such a practice serves to diminish the image of companies like the Maersk Group as a strong global representative of green recycling. Better examples of green recycling efforts for end-of-life vessels are from Norway, the Netherlands, and Canada. In Norway, the shipping firms Grieg and Hoegh have been stalwarts in setting positive examples for best practices in ship recycling and have no record of beaching vessels. Shipping firms like Canada Steamship Lines and Royal Dutch Boskalis have extended their environmental consciousness further by limiting their recycling to countries that are part of the Organization for Economic Co-operation and Development (OECD). The mission of the OECD is to promote policies that improve the economic and social well-being of people around the world. That includes providing a forum in which governments work together to share experiences and seek solutions to environmental problems. The organization works with governments to understand environmental threats and to set international standards on the safety of chemicals and chemical exposure. Since March 1992, transboundary movements of wastes destined for recovery operations between member countries of the OECD have been supervised and controlled under a specific intra-OECD Control System that facilitates the trade of recyclables in an environmentally sound manner by using a simplified procedure as well as a risk-based approach to assess the necessary level of control for materials. Wastes exported outside the OECD area, whether for recovery or final disposal, do not benefit from this simplified control procedure (European Commission, 2016).

From a date set in the regulation between mid-2017 and December 31, 2018, large commercial seagoing EU vessels are required to be recycled in "safe and sound" ship recycling facilities included in the European List of ship recycling facilities. This list was first established on December 19, 2016. Any ship recycling facility, regardless of its location, must comply with multiple safety and environmental requirements to be included in the European List. The European Commission issued technical guidelines on these requirements in April of 2016. For Third World countries, the European Commission is responsible for conducting reviews of applications received from the ship recycling facilities. It is a bit different for facilities located in EU nations. There, the national authorities of the Member States are held responsible for notifying the Commission which facilities are compliant (European Commission, 2016).

Under the revised Ship Recycling Regulation, the installation or use of certain hazardous materials on ships are either totally prohibited or, at least, restricted. Asbestos and ozone-depleting substances are among the hazardous materials included. A key requirement of the regulation is that every new European ship (or a ship flying the flag of a third country at an EU anchorage) possess an on-board inventory of hazardous materials, verified by the relevant authority that details the location and amounts of the materials. In addition, the port authorities of EU Member States were authorized to monitor European ships to substantiate if they have on board ready-for-recycling certificates or a valid inventory of hazardous materials. In November 2016, EMSA, the European Maritime Safety Agency, published a Best Practice Guidance on the Inventory of Hazardous Materials for practitioners in the field, ship owners, and national authorities (European Commission, 2016).

Despite the noble efforts of relevant international authorities and agencies, illegal ship breaking continues to be a preferred method of environmental offenders desiring to avoid the cost of properly handling end-of-life vessels. In June of 2017, The Brazilian CONTTMAF trade union federation singled out the ship-breaking practices of Transpetro. Transpetro is an oil and gas transportation subsidiary of the Brazilian petroleum corporation Petrobras and is responsible for dismantling more than twenty vessels on the beaches of India and Pakistan. In addition, based on data from the NGO Shipbreaking Platform, another Brazilian corporation, Vale, sent five ships to ship-breaking beaches in Bangladesh and Pakistan between 2015 and 2017, where at least 70 workers were severely injured in dismantling operations. Thus, it is clear that well-intentioned strengthening of regulations addressing ship breaking has not completely eradicated the unethical and criminal actions of ship-breaking offenders (NGO Shipping Platform, June 15, 2017).

Discussion Questions

1) This chapter discusses the general circumstances in which environmental crimes are committed as well as methods employed to commit them. Discuss how, in some ways, the circumstances and methods of environmental crime have remained the same over time and how, in other ways, these circumstances and methods have changed. Why do you think this is so?
2) What impact do you think the elements of Adler's environmental crime triangle have on the characteristics of environmental crimes from past decades? For the more present? Use the Adler environmental crime triangle to support your projection of how environmental crime will look in the future.

References

Adler, F. (1996). Offender-specific vs. offense-specific approaches to the study of environmental crime. In S. Edwards, T. Edwards, and C. Fields (Eds.), *Environmental crime and criminality: Theoretical and practical issues*. New York: Garland Publishing Inc.

Caniglia, J. (2014, March 20). Fracking employee sentenced to probation for dumping thousands of gallons of waste into Mahoning River. *The Plain Dealer.*

Cohen, L., & Felson, H. (1979). Social change and crime rate trends: A routine activity approach. *American Sociological Review, 44*, 588-608.

European Commission. (2016, December 12). Ship recycling. Retrieved from http://ec.europa.eu/environment/waste/ships/

Goodnough, A. (2016, August 30). Their soil toxic, 1,100 Indiana residents scramble to find new homes. *The New York Times.* Retrieved from http://www.nytimes.com/2016/08/31/us/lead-contamination-public-housing-east-chicago-indiana.html?_r=0

Hood, G. (2016, August 4). One year after a toxic river spill, no clear plan to clean up western mines. All things considered. *National Public Radio.* Retrieved from http://www.npr.org/2016/08/04/488579040/one-year-after-a-toxic-river-spill-no-clear-plan-to-clean-up-western-mines

Intellihub (2014, September 9). New York property owner and manager sentenced to 21 months in federal prison for conspiring to violate the Clean Air Act. https://www.intellihub.com/new-york-property-owner-manager-sentenced-21-months-federal-prison-conspiring-violate-clean-air-act/

Mueller, G. (1996). An essay on environmental criminality. In S. Edwards, T. Edwards, and C. Fields (Eds.), *Environmental crime and criminality: Theoretical and practical issues.* New York: Garland Publishing Inc.

NGO Shipping Platform. (2014, February 3). *NGOs publish 2013 list of toxic ship dumpers: German and Greek companies amongst the worst.* Brussels. Retrieved from http://www.shipbreakingplatform.org/press-release-ngos-publish-2013-list-of-toxic-ship-dumpers-german-and-greek-shipping-companies-amongst-the-worlds-worst/

NGO Shipping Platform. (2017, June 15). *Brazilian government asked to stop dumping toxic ships on South Asian Beaches.* Brussels/Rio de Janeiro. Retrieved from http://www.shipbreakingplatform.org/press-release-brazilian-government-asked-to-stop-dumping-toxic-ships-on-south-asian-beaches/

Rebovich, D. (2015). *Dangerous ground: The world of hazardous waste crime.* New Brunswick, NJ: Transaction Publishers.

The Economist. (2015, August 15). *Arsenic and old face: The agency charged with protecting the environment pours poison on it*, vol. 416, no. 8975. Washington, DC, p. 23.

U.S. DOJ. (2013, August 2). Two sentenced in New York State for dumping thousands of tons of asbestos in violation of the clean water act. Retrieved from https://www.justice.gov/opa/pr/two-sentenced-new-york-state-dumping-thousands-tons-asbestos-violation-clean-water-act

U.S. DOJ. (2015, January 23). Owners & managers of former salvage operations at former textile Plant in Tennessee sentenced to prison for conspiracy associated with illegal asbestos removal. Retrieved from https://www.justice.gov/opa/pr/owners-managers-former-salvage-operations-former-textile-plant-tennessee-sentenced-prison

U.S. DOJ. (2015, November 19). *Offshore oil platform operator agrees to more than $41 million in penalties for unauthorized oil discharges and improper operations in Gulf of Mexico.* Retrieved from https://www.justice.gov/opa/pr/offshore-oil-platform-operator-agrees-more-41-million-penalties-unauthorized-oil-discharges

U.S. DOJ. (2015, December 3). General Electric to pay $2.25 million for violating federal and state environmental laws in Waterford, New York. Retrieved from https://www.justice.gov/opa/pr/general-electric-pay-225-million-violating-federal-and-state-environmental-laws-waterford-new

U.S. DOJ. (2015, December 11). Engineering officers charged in scheme to cover up oil discharges from cargo vessel. Retrieved from https://www.justice.gov/opa/pr/engineering-officers-charged-scheme-cover-oil-discharges-cargo-vessel

U.S. DOJ. (2016). *Prosecution of federal pollution crimes.* Washington, DC: Environmental Crime Section, U.S. Department of Justice.

U.S. DOJ. (2016, February 18). Ship captain pleads guilty to felony obstruction related to pollution from tanker ship traveling to Charleston. Retrieved from https://www.justice.gov/opa/pr/ship-captain-pleads-guilty-felony-obstruction-related-pollution-tanker-ship-traveling

U.S. DOJ. (2016, April 8). Norwegian shipping company sentenced in Alabama to pay $2.5 million for illegally discharging oil into the ocean. Retrieved from https://www.justice.gov/opa/pr/norwegian-shipping-company-sentenced-alabama-pay-25-million-illegally-discharging-oil-ocean

U.S. DOJ. (2016, August 9). Justice Department, EPA and the state of New Mexico announce settlement for $143 million cleanup at the chevron questa mine. Retrieved from https://www.justice.gov/opa/pr/justice-department-epa-and-state-new-mexico-announce-settlement-143-million-cleanup-chevron

U.S. EPA. (2016). Learn the issues. Asbestos NESHAP. Retrieved from https://www.epa.gov/asbestos/asbestos-neshap.

U.S. FBI. (2014, May 13). U.S. Attorney's Office announces final conviction involving Kensington towers asbestos project. Retrieved from https://www.fbi.gov/contact-us/field-offices/buffalo/news/press-releases/u.s.-attorneys-office-announces-final-conviction-involving-kensington-towers-asbestos-project

Wattles, J., & Egan, M. (2016, September 3). Oklahoma orders shutdown of 37 wells after earthquake. *CNNMoney*. Retrieved from http://money.cnn.com/2016/09/03/news/economy/oklahoma-earthquake-fracking-oil/

WBNG News. (2015, May 13). *Owego man pleads guilty to environmental crime*. Retrieved from http://www.wbng.com/news/local/Owego-man-pleads-guilty-to-environmental-crime-303658661.html

Chapter 7

Who Protects Us from Environmental Crime and How Effective Are They?

Introduction

Chapter 7 describes those government agencies and private entities that protect society from criminal conduct that is harmful to the environment.

Environmental protection occurs in two ways: proactively (through prevention) and reactively (through enforcement, i.e., by detection, investigation, and prosecution). Proactive protection efforts are maintained by government agencies, non-governmental organizations (NGOs), and businesses. Enforcement efforts are primarily the responsibility of governmental agencies. We will see, however, that occasionally the activities of NGOs will result in governmental enforcement.

This chapter approaches the discussion of environmental crime protection by separately considering those proactive and reactive activities designed to ensure a safe environment through the identification of the agencies and organizations that are involved in each method of protection and an illustration of the activities of each that are designed to ensure a safe environment.

Proactive Prevention of Environmental Crime

Government Regulation

Environmental Protection Agency (EPA)

The EPA was created in 1970 but not by legislation. Instead, the agency was created by an executive order signed by President Richard Nixon on December 4, 1970. The mission of the agency is to protect human health and the environment.

Aside from EPA's responsibility to enforce numerous environmental laws and regulations, the agency is also tasked with the prevention of environmental harm through proactive measures. For example, the EPA has produced guidance for schools on measures the schools can deploy to prevent air quality issues from arising. EPA provides seminars and training sessions on topics within its regulatory jurisdiction. Most notably, the EPA Office of Research and Development conducts research designed to prevent future environmental harm, and the agency funds research and training of state environmental agencies in proactive measures that can be employed to minimize environmental harm.

Department of the Interior

The U.S. Department of the Interior was created by Congress on March 3, 1849.

The current environment-related responsibilities of the Department include land and resource management (including National Parks), wildlife conservation, surface mining reclamation, and offshore drilling management. The greater part of its effort is research and activities designed to prevent harm to wildlife and the environment of federal lands.

Non-Governmental Organization Strategies to Protect the Environment

There are numerous organizations that focus their efforts on the protection of the environment. Though not possible to consider the efforts of every organization here, some of the more notable are the Sierra Club, Environmental Defense Fund, and Greenpeace.

Sierra Club

Founded in 1892 in California, the Sierra Club boasts some 3.5 million members and supporters. It is considered the largest and most influential of the grassroots environmental organizations. Sierra Club lobbies extensively before state and federal legislatures and often engages in litigation to prevent activities it views as a danger to the environment.

Environmental Defense Fund

The Environmental Defense Fund was created in 1967. It began as a legal effort to ban the use of the pesticide DDT. Building on that success, the Fund, a national non-profit organization, has grown to some two million members. Much of the Fund's work consists of economics and market research and publications, as well as partnerships with industry designed to address specific environmental issues.

Greenpeace

Greenpeace was founded in 1971. Far more global in its reach, Greenpeace claims some 2.8 million members worldwide.

Greenpeace is far more activist in its support of environmental causes and in its opposition to activities it deems harmful to the environment. It conducts lobbyist and educational activities in many countries.

Reactive Enforcement and Prosecution of Environmental Crime

Although federal and state laws provide for proactive measures to protect the environment, the reactive measures of enforcement and prosecution garner the greatest attention.

Federal Government

Environmental Protection Agency

The Environmental Protection Agency can undertake three types of enforcement measures: (1) a civil administrative proceeding; (2) a formal civil action; and (3) criminal proceeding.

Civil administrative proceedings are in-house (within the agency) proceedings often initiated when a violation of a regulation or law is first discovered and typically are designed to urge the offender into compliance. Examples of such proceedings would be a Notice of Violation and Cease and Desist Order, or an agency order directing an offender to clean up the damages caused by a violation. If the matter is contested by the alleged offender, the agency may conduct a hearing before an administrative hearing officer and issue a determination that could include the imposition of a civil penalty.

A civil action is a more formal judicial procedure that now may involve a trial in federal district court. This alternative often is utilized when the alleged offender failed to comply with an administrative order; failed to comply with, or pay for, a cleanup order; or violated an environmental law or regulation. If found responsible, the offender usually is required to pay damages and is enjoined by the court to do, or to refrain from doing, certain activity.

The agency may also conduct a criminal investigation. This alternative is often sought where the alleged offender has committed several, or repeated, violations of laws or regulations, or where the violation is serious and the offender has acted willfully. In these cases, the EPA will undertake an investigation by its special agents in the Criminal Investigation Division. If the offense is considered to warrant a criminal action, the Administrator of the EPA will refer the matter to the United States Department of Justice (USDOJ) for criminal action. At that point, the decision as to whether a criminal proceeding should be initiated will be made by the USDOJ. In the event that the USDOJ decides against a criminal prosecution, the agency may still proceed with a civil administrative proceeding or a formal civil action.

Special agents of the Criminal Investigation Division (CID) are located in more than 40 of the EPA's resident offices, thereby enabling the agency to respond to complaints or incidents quickly. These agents are sworn law enforcement officers and have full criminal investigative and enforcement powers on environmental matters and on any other federal crime (see 18 U.S.C. §3063). The EPA currently has 147 CID special agents, well below the minimum contingent of 200 agents authorized by Congress in 1990 (https://www.govexec.com/management/2017/08/epa-has-slashed-its-criminal-investigation-division-half/140509/).

Department of the Interior (DOI)

The Department of the Interior was created by Congress in 1849. The Department is composed of ten Bureaus, including the Bureau of Land Management, Bureau of Safety and Environmental Enforcement, Bureau of Reclamation, National Park Service, Fish and Wildlife Service, and the Office of Surface Mining Reclamation and Enforcement. It thus has the responsibility for several areas associated with the environment.

The Bureau of Safety and Environmental Enforcement (BSEE) is the part of the DOI closely associated with environmental protection. The Bureau's mission is to

> promote safety, protect the environment and conserve resources offshore through rigorous regulatory oversight and enforcement (www.bsee.gov/public-site-page/bsee-strategic-plan).

It thus oversees the protection of federal lands and offshore development. The Bureau engages in inspections and investigations, but its investigators are not law enforcement personnel.

The Fish and Wildlife Service, however, includes an Office of Law Enforcement that employs some 250 Criminal Investigation Division special agents. Like their counterparts in the EPA, these special agents have full law enforcement powers at the federal level and occasionally collaborate with EPA agents on certain investigations. Their primary responsibility, though, involves criminal conduct that affects fish and wildlife, such as chemical spills, illegal use of pesticides, and wildlife smuggling (www.fws.gov/le/special-agents.html).

Department of Justice

The United States Department of Justice is the federal government agency that is responsible for all criminal prosecutions on behalf of the United States and its various agencies. Thus, all prosecutions for environmental crime fall within the purview of this agency.

The Environment and Natural Resources Division of USDOJ is the nation's environmental lawyer team. The Division is responsible for the initiation of civil and criminal proceedings involving the violation of laws within the jurisdiction of the EPA and the Department of the Interior.

The Environment and State and Local Governments

Every state has an agency responsible for the enforcement of its environmental laws and federal mandates.

New York: Department of Environmental Conservation (NYSDEC)

New York's Department of Environmental Conservation was created on July 1, 1970. Its mission is to

> To conserve, improve and protect New York's natural resources and environment and to prevent, abate and control water, land and air pollution, in order to enhance the health, safety and welfare of the people of the state and their overall economic and social well-being.

As illustrated on its website (www.dec.ny.gov), the DEC performs both proactive (management of forest areas, regulation of fishing and hunting, conservation of natural resources) and reactive (enforcement of criminal environmental laws and regulations) efforts. On the enforcement side, the agency has more than 330 sworn Environmental Conservation Police Officers (ECOs) that possess full police powers to enforce environmental crimes and other state crimes. The ECOs focus on two enforcement areas: fish and wildlife (hunters, trappers, commercial fishermen) and environmental quality (water pollution, excessive commercial vehicle emissions, improper use of pesticides, illegal mining). That law enforcement division also includes the Bureau of Environmental Crimes Investigation, the detective squad for environmental crimes enforcement.

California: Environmental Protection Agency (CalEPA)

The California Environmental Protection Agency was created as a cabinet-level agency in 1991. Its mission is:

to restore, protect and enhance the environment, to ensure public health, environmental quality and economic vitality.

CalEPA is organizationally divided into six areas: pesticide regulation, water resources, air resources, recycling, toxic substances, and environmental health hazards. Each area employs enforcement officers and inspectors. It is noted that CalEPA has two program initiatives that are regularly funded: the Environmental Justice Task Force and the Environmental Circuit Prosecutor Project. The latter project provides specialized prosecutors to rural counties that otherwise might lack the expertise or resources to handle environmental crimes committed in their area.

The official website of CalEPA (https://calepa.ca.gov/enforcement) lists the numerous environmental crime cases pursued by the agency over the past several years. CalEPA, though, is perhaps the nation's leader in proactive steps designed to protect the environment. The most recent publication on accomplishments, *Major Accomplishments: 2011–2018*, documents numerous proactive measures on climate control, pesticide control, and the protection of water and air resources.

Florida: Department of Environmental Protection (FDEP)

The Florida Department of Environmental Protection was created in 1993. Its mission:

> protects, conserves and manages the state's natural resources and enforces its environmental laws.

Its vision is:

> to create strong community partnerships, safeguard Florida's natural resources and enhance its ecosystems.

The FDEP is divided into three focus areas: land and management, which provides for the acquisition, maintenance, and restoration of parks and recreation areas; regulatory, which oversees the permit process and compliance of air and water quality and manages waste cleanups; and ecosystem restoration programs for the protection and improvement of aquatic resources such as the Everglades, springs, and coastal areas (https://floridadep.gov/about-dep). The enforcement of environmental laws and regulations is provided by the Department's six districts, organized geographically. Each district is responsible for the review of permit applications, inspecting facilities that have permits, conducting compliance assessments, and investigating criminal and civil violations. FDEP employs less than 3,000 people, about 155 fewer than in 2011. However, employment is expected to increase as a result of the current budget approved by Florida's Governor.

In 1998, the state of Florida created the Florida Fish and Wildlife Conservation Commission (FWC) and invested the Commission with the responsibility to manage the recreational parks as well as fish and wildlife resources. At that time, FDEP's sworn law enforcement officers were employed with the Division of Law Enforcement within the Bureau of Park Patrol. The Bureau of Park Patrol was merged into the FWC, and the Division of Law Enforcement is now part of the FWC. Any remaining FDEP sworn officers would now be employees of the FDEP's Division of Emergency Management, which provides emergency response to hazardous waste spills and similar emergencies. Thus, those FDEP employees responsible for the investigation of air and water pollution incidents are not sworn law enforcement officers.

FDEP's South District has actively monitored compliance by the City of Fort Myers to remove lime sludge from a city neighborhood, which is the subject of a case study in Chapter 8.

West Virginia: Department of Environmental Protection (WVDEP)

The West Virginia Department of Environmental Protection was created in October 1991. Its mission is simply stated: to promote a healthy environment. Programmatically, WVDEP is organized in three primary areas: the Division of Air Quality, Division of Mining and Land Reclamation, and the Division of Water and Waste Management.

Separate enforcement units exist within each division. The Section of Compliance and Enforcement in the Division of Air Quality is responsible for the inspection and investigation of air pollution sources, monitoring asbestos removal and demolition, and implementing federal requirements pertaining to air quality (https://dep.wv.gov/daq/CandE/Pages/ default.aspx). The Division of Mining and Land Reclamation has an Inspector and Enforcement unit that routinely inspects mines and reclamation projects. Inspectors have the authority to initiate civil or criminal investigations and to assist in criminal prosecutions (https://dep.wv.gov/dmr/ handbooks/ Documents/Inspection%20and%20Enforcement%20Handbook/Section%201.pdf). The Office of Environmental Enforcement is housed within the Division of Water and Waste Management. It is responsible for ensuring compliance with West Virginia's Solid Waste Management Act, Water Pollution Control Act, Groundwater Protection Act, Hazardous Waste Management Act, Underground Storage Tank Act, and the Dam Safety Act (https://dep.wv.gov/WWE/ee/Pages/ default.aspx). The inspection and investigation of the discharge of materials into state waters is the subject of the case study discussed below.

The environment protection agencies of each state focus, or prioritize, on different areas based on different geographic, geologic, and environmental issues in the state. For example, West Virginia has a distinct focus on mining, an area that would not be prominent in Florida or New York. California's emphasis on air quality is as prominent there as water quality in Florida.

The states also are mandated by the EPA to provide air quality and water quality standards and regulations and to cooperate in the enforcement of those standards and regulations. The following case study is one illustration of cooperation between enforcement agencies of two states and the federal government concerning illegal discharges during fracking operations.

Case Study: Fracking and Federal-State Enforcement Cooperation

The federal Clean Water Act (33 U.S.C. §§1251–1387) prohibits the discharge of fill and waste materials into waters of the United States without obtaining a permit from the U.S. Army Corps of Engineers. The Pennsylvania Clean Streams Law contains a similar provision requiring a permit. Sections 3, 6, and 8 of West Virginia's Water Pollution Control Act have a similar requirement.

In 2010, an inspector employed by the Pennsylvania Department of Environmental Protection (PADEP) conducted an unannounced inspection of a hydraulic fracking site (the Marquardt site) in Hughesville, Penn Township, Lycoming County, Pennsylvania, operated by XTO Energy, Inc., a subsidiary of ExxonMobil. That inspection revealed the discharge of gas well wastewater and other materials from storage tanks into a tributary of Sugar Run and Sugar Run itself, which is a tributary of the Susquehanna River, in violation of the Pennsylvania Clean Streams Law and Solid Waste Management Act. Further investigation by PADEP led to criminal charges filed in 2013 by the Pennsylvania Attorney General's Office for "illegally discharging more than 50,000 gallons of toxic waste water" (www. processingmagazine.com/pennsylvania-files-charges-against-xto-energy-over-fracking-wastewater-spill/). While PADEP conducted its investigation, the EPA also investigated and ultimately brought a civil action against XTO Energy for violations of the Clean Water Act (33 U.S.C. §1311). The federal proceeding was terminated when the United States entered into a consent order with XTO Energy, Inc. (*United States v XTO Energy, Inc.*, Notice of Proposed Consent Decree [2013]). Under that consent order, XTO Energy, Inc., agreed to pay a civil penalty of $100,000. The state of Pennsylvania subsequently also agreed to discontinue criminal proceedings under an agreement that required XTO Energy to pay a fine of $300,000 to PADEP and an additional $100,000 to the Susquehanna Greenway Partnership for the costs of a cleanup of the spill (https://stateimpact.npr.org/pennsylvania/2016/08/04/attorney-general-reaches-agreement-with-xto-over-criminal-charges/).

In 2011, the West Virginia Department of Environmental Protection (WVDEP) and the U.S. Army Corps of Engineers discovered through joint inspections that XTO Energy had discharged waste material, some of it toxic, into waterways at three different sites in northern West Virginia. Initially, the inspections revealed violations at three sites. XTO Energy then conducted an audit of its operations, and that audit revealed violations at additional sites. Finally, joint federal and state inspections by the EPA, U.S. Army Corps of Engineers, and WVDEP disclosed violations at three additional sites. In all, the discharges affected 5,300 linear feet of streams that are tributaries of the Monongahela River and 3.38 acres of wetlands on eight sites in three counties. The EPA charged XTO Energy in a civil complaint with violations of sections 301 and 404 of the Clean Water Act, and West Virginia also initiated a civil action for violations of state law. The civil proceedings ultimately were discontinued when the state and the United States entered into a consent decree requiring XTO Energy to pay a civil penalty of $2.3 million, with 50% being paid to the state of West Virginia and 50% to the United States government.

The successful collaborative enforcement of Pennsylvania, West Virginia, and federal agencies has been featured in seminars presented by those agencies.

Discussion Questions

1) Florida's law enforcement personnel with training on environmental issues are employed by Florida's Fish and Wildlife Conservation Commission and not by its Department of Environmental Protection. Does it seem that this organization alignment would have an impact on the criminal enforcement of standards and requirements pertaining to air and water quality?
2) The Pennsylvania and West Virginia efforts in collaboration with the EPA and U.S. Army Corps of Engineers regarding the protection of water quality is but one example of ways that government agencies cooperate in their efforts to protect the environment. Discuss other avenues or approaches that are available to such government agencies.

Bibliography

California Environmental Protection Agency (CalEPA). (2018). *Major accomplishments: 2011–2018*. Last retrieved on June 24, 2019 at https://calepa.ca.gov/wp-content/uploads/sites/6/2019/03/CalEPA_Accomplishments_Report_2011-2018_a.pdf

Processing Magazine. (2013). *Pennsylvania charges against XTO energy over fracking wastewater spill*. Last retrieved on June 24, 2019 at https://www.processingmagazine.com/pennsylvania-files-charges-against-xto-energy-over-fracking-wastewater-spill/

United States Department of Justice (USDOJ). (2017, November). Monthly Bulletin, Environmental Crimes Section. Retrieved from https://www.justice.gov/sites/default/files/enrd/legacy/2015/04/13/monthly-bulletin-july2012_508.pdf

United States Department of Justice (USDOJ). (2014). *XTO Energy Inc. to restore areas damaged by natural gas extraction activities*. Last retrieved on June 24, 2019 at https://www.justice.gov/opa/pr/xto-energy-inc-restore-areas-damaged-natural-gas-extraction-activities

United States Environmental Protection Agency (EPA). (2014). *XTO Energy, Inc. To restore areas damaged by natural gas extraction activities*. Last retrieved on June 24, 2019 at https://www.epa.gov/enforcement/reference-news-release-xto-energy-inc-restore-areas-damaged-natural-gas-extraction

United States v XTO Energy Inc. (M.D. Pa. 2013). *Consent decree*. Last retrieved on June 24, 2019 at http://www.velaw.com/UploadedFiles/VEsite/Presentations/ENV_ENFORCEMENT-2381522-v1-XTO_Lodged_Decree.pdf

United States v XTO Energy Inc. (N.D. WVa. 2014). *Complaint*. Last retrieved on June 24, 2019 at https://www.justice.gov/file/189056/

United States v XTO Energy Inc. (N.D. WVa. 2014). *Notice of consent decree*. Last retrieved on June 24, 2019 at https://www.justice.gov/file/189061/

West Virginia Department of Environmental Protection. (2019). *Agency history*. Last retrieved on June 24, 2019 at https://dep.wv.gov/Pages/OrganizationalHistoryandOverview.aspx

Chapter 8

Justice for All? Are We Achieving Environmental Justice in the United States?

Introduction

In this chapter, we consider the concept of environmental justice and its relationship to environmental crime. We begin by considering the meaning of the term "environmental justice" and related terms, such as "environmental racism" and "environmental equality," and then discuss the various types of conduct that can potentially lead to issues of environmental justice, and finally the relationship of environmental justice and environmental crime. We then present two case studies that illustrate the injustice that can result from environmental policy and misconduct.

Environmental Justice Defined

Environmental justice, as a concept or movement, began in the 1980s. The catalyst was the siting of a landfill in Warren County, North Carolina, for the dumping of 60,000 cubic yards of soil contaminated with toxic waste consisting of polychlorinated biphenyl (PCB)-laden oil.

Although the concept of environmental justice has been defined in different ways, the most widely accepted definition is that set forth in 1998 by the U.S. Department of Environmental Protection:

> the fair treatment and meaningful involvement of all people regardless of race, color, national origin or income with respect to the development, implementation, and enforcement of environmental laws, regulations and policies.

Various words and phrases within this definition have important meanings. We explore this aspect later in the chapter.

Environmental Racism, Environmental Equity, and Related Terms

Environmental Justice is the more encompassing of the various terms used to describe, in one way or another, the concept of environmental justice. Other terms include environmental racism, environmental equality, and environmental equity.

Environmental Racism

Although environmental racism is viewed as a subset of the more encompassing environmental justice, it was environmental racism that led to the environmental justice movement, as illustrated by the siting of the Warren County landfill. In fact, one study concerning the siting of toxic waste landfills in southern states concluded that race "proved to be the most significant among variables tested in association with the location of commercial hazardous waste facilities" (UCC, p. xiii).

It has been suggested that the term environmental racism was "coined" by Dr. Benjamin Chavis, Jr., during Congressional testimony offered in 1982 (Frye, 1993). In his testimony, Dr. Chavis defined it as

> racial discrimination in environmental policy making and the unequal enforcement of environmental laws and regulations. It is the deliberate targeting of people of color communities for toxic waste facilities and the official sanctioning of a life threatening presence of poisons and pollutants in people of color communities. It is also manifested in the history of excluding people of color from the leadership of the environmental movement. (Frye, p. 56)

Robert Bullard (1994b, p. 451), considered by many as the Father of the environmental justice movement has defined environmental racism as:

> any policy, practice, or directive that differentially affects or disadvantages (whether intended or unintended) individuals, groups, or communities based on race or color.

Bullard (1994b, pp. 451–452) further explains:

> Environmental racism combines with public policies and industry practices to provide benefits for whites while shifting losses to people of color. Environmental racism is reinforced by government, legal, economic, political, and military institutions.
>
> Many environmental decisions distribute the costs in a regressive pattern, while providing disproportionate benefits for individuals who fall at the upper end of the socioeconomic spectrum. In the United States, race has been found to be independent of class in the exposure to lead, harmful pesticides, location of municipal landfills and incinerators, abandoned toxic waste dumps, and environmental protection and cleanup of Superfund sites.
>
> Professor Bunyan Bryant (1995, p. 5) similarly noted that environmental racism
>
> refers to those institutional rules, regulations, and policies or government or corporate decisions that deliberately target certain communities for least desirable land uses, resulting in the disproportionate exposure of toxic and hazardous waste on communities based upon certain prescribed biological characteristics.

Although race continues to be a major factor in environmental policies, decisions, and enforcement, the term more commonly employed today is environmental justice.

Environmental Equity

Attempts to define the term "environmental equity" have not been consistent.

Some have limited the term to "equal protection of environmental laws" (Bryant, 1995, p. 5). Others have considered the term to relate to the general notion of fairness (e.g., Bullard, 1994d, p. 12), and yet others have considered the term to relate to economic and risk assessment concerns. For example, locating hazardous waste sites in low-income neighborhoods and particularly in rural communities is viewed as reducing the risk of environmental harm and more "equitable" because there is a promise of economic opportunities in the form of jobs for those in the community.

Environmental Equality

Perhaps the terminology least used is environmental equality. Many references to environmental equality are discussions in response to the 2004 work of David Pellow, entitled *Garbage Wars: The Struggle for Environmental Justice in Chicago*. Pellow viewed environmental equality as "focusing on whether or not unequal outcomes exist before addressing the causes of environmental inequality" (Lynch et al., 2014, p. 252).

The Development of the Environmental Justice Movement

In theory, environmental justice is considered to be the result of a melding of two movements: social justice and civil rights with environmental law and policy (Heagerty, 2010).

Although some suggest that the environmental justice movement began with the enactment of the Civil Rights Act of 1964 (Carter, n.d.), the consensus is that the movement toward environmental justice began with the Warren County, North Carolina, protests against the siting of a landfill for toxic waste (PCB-laden soil) in the late 1970s and early 1980s (Cole, 2007). The first notable legal action brought to prevent the location of a solid waste landfill in a minority community occurred in Houston, Texas, in 1979 (Bean v Southwestern Waste Management Corp., 482 F. Supp. 673 [S.D. Tex. 1979]), and other legal, legislative, and environmental events occurred prior to the Warren County protests, including a lawsuit that ultimately resulted in the banning of the DDT pesticide, the 1971 report of the President's Council on Environmental Quality that recognized the adverse effect of environmental policies upon minorities and the poor particularly in urban areas, and measures taken to prevent lead poisoning of children in urban areas. However, it was the 1982 opposition to the Warren County toxic dump that garnered national media coverage and signified the engagement of civil rights forces in environmental policies and actions.

Following the protests in Warren County, Congressman Fauntroy, himself a protestor there, requested that the U.S. General Accounting Office (GAO) study the correlation between the location of hazardous waste landfills and the racial and economic status of the communities surrounding those landfills. The GAO issued its report in June 1983, confirming what many had believed; of the four hazardous waste landfills in the eight southern states comprising Region IV of the Environmental Protection Agency (EPA), African-Americans comprised more than a majority of the nearby population in three of those locations. In one instance, African-Americans constituted

90% of the affected population. Further, at least 26% of the population in all four locations lived below the poverty level, and of that population, at least 90% were African-American.

Four years later, the Commission for Racial Justice of the United Church of Christ (UCC) issued a report that considered the correlation between race and socio-economic characteristics and the siting of hazardous waste facilities throughout the nation. The study included the location of commercial hazardous waste sites as well as uncontrolled sites; the latter category is an EPA classification consisting of sites that had been closed or abandoned or that resulted from accidental spills or illegal discharges. The study determined that race was the most significant factor in the location of hazardous waste sites, finding that three out of five African-Americans and Hispanics lived in communities with uncontrolled waste sites. The UCC Commission advanced numerous recommendations, including that the President of the United States issue an executive order directing federal agencies to consider the impact of current policies and regulations on racial and ethnic communities and that the EPA create an office to specifically ensure that racial and ethnic concerns be considered in relation to hazardous wastes and in the cleanup of hazardous waste sites (UCC, 1987).

It would not be until 1994 that the environmental justice movement gained its first major success. President Clinton issued Executive Order 12,898 (3 C.F.R. §859 [1995]), which significantly provided:

> Each Federal agency shall conduct its programs requiring federal agencies to develop an environmental justice strategy.

The Order required each federal agency, to the extent practical, to make achieving environmental justice part of the agency's mission "by identifying and addressing, as appropriate, disproportionately high and adverse human health or environmental effects of its programs, policies, and activities on minority populations and low-income populations." Additionally, the Order required each agency to:

> conduct its programs, policies, and activities that substantially affect human health or the environment, in a manner that ensures that such programs, policies, and activities do not have the effect of excluding persons (including populations) from participation in, denying persons (including populations) the benefits of, or subjecting persons (including populations) to discrimination under, such programs, policies, and activities, because of their race, color, or national origin.

The Order also requires the Director of the EPA to convene an interagency Federal Working Group on Environmental Justice and requires each agency to develop an environmental justice strategy that identifies policies, practices, and activities disproportionately affecting minority and low-income populations, and that includes revisions designed to:

> (1) promote enforcement of all health and environmental statutes in areas with minority populations and low-income populations; (2) ensure greater public participation; (3) improve research and data collection relating to the health of and environment of minority populations and low-income populations; and (4) identify differential patterns of consumption of natural resources among minority populations and low-income populations.

Soon after issuance of the Order, the EPA, under the leadership of its then Administrator, Carol Browner, issued its first environmental justice strategy (Pub. No. EPA-200-R-95-002). Significantly, the strategy considered a "major priority" changes in the permitting process designed to diminish discrimination in the siting of hazardous waste sites in minority and low-income communities, as well as ensuring increased public participation in the siting decision process. Other federal agencies are required to submit an environmental justice strategy and to provide an annual progress report regarding its accomplishments in achieving environmental justice. The environmental justice movement had achieved government recognition and support. Whether the goals of environmental justice have been achieved is discussed later in this chapter.

The Meaning of Environmental Justice

Executive Order 12,898 did not define "environmental justice." The EPA has, however, developed a definition (provided at the beginning of the chapter) that has met with wide acceptance. It is important to consider the importance of two terms within the definition: what they mean and don't mean.

"Fair Treatment"

The quest for environmental justice seeks "fair" treatment. Fair treatment implicitly does not mean "equal" treatment. In the context of locating a hazardous waste site, for example, typically there will only be one site at issue. So, it is not a matter of equal treatment because that suggests two sites or two communities or populations. Even if the first waste site was located in a disproportionate minority population, would equal treatment necessarily mean locating a second site in an equally disproportionate non-minority population? Clearly not; even if the first site was in a community of 60% minority population, it might well be impossible to find an environmentally sound location with a 60% or more non-minority population. Thus, environmental justice means fair treatment.

"Meaningful Involvement"

Likewise, the goal is to provide meaningful involvement of minority or low-income populations in the development of environmental policies and decisions as well as the implementation of enforcement proceedings. In many of the location determinations for hazardous waste sites in southern states during 1960–1990, the minority population in each community lacked knowledge of the prospective siting or membership in political bodies and thus had no opportunity for involvement in the location decision (Bullard, 1994a; UCC, 1987). One of the recommendations urged by various studies and activist groups was greater participation of minority and low-income groups in the political/environmental decision made concerning site location (e.g., UCC, 1987). Meaningful involvement does not mean that a determination concerning the location of a hazardous waste site cannot be made unless local executives and legislatures appoint persons representing minority and low-income communities to political bodies (planning and zoning boards, for example) that make those decisions. It does mean, however, that those political bodies must ensure that the affected communities have meaningful input into the decision process. The U.S. Commission on Civil Rights has recommended that "[f]ederal agencies should require state and local zoning and land-use authorities, as a condition to receiving and continuing to receive federal funding, to incorporate and implement the principles of environmental justice into their zoning and land-use policies" (Commission, 2003, p. 28). A good sense of the essence of meaningful involvement can be gleaned from a review of recommendations made by the Commission at pp. 170–171 of its 2003 report.

The Types of Actions That Lead to Claims of Environmental Injustice

The Siting of Waste Facilities

By far, the main type of action leading to claims of environmental injustice or environmental racism is the siting of waste landfills (many of them toxic waste) or repositories. The 1983 GAO report notes that there were then four toxic waste landfills in the south and that three of those sites were located where the population consists of a majority of African-Americans. The largest of those sites (in fact, the largest toxic waste landfill in the country) is in Emelle, Alabama, a rural area of Sumter County in western Alabama on the Mississippi border. Ninety percent of the local population was African-American, and 42% of the population lived below the poverty level, all of whom were African-American.

Bullard (1994a, p. 60) notes that the Emelle facility operated by Chemical Waste Management (Chemwaste) was "forced on the people." No African-Americans held public office or sat on any governing body that would have been able to provide input regarding the siting of the facility. Based upon a local newspaper account, local residents thought that the new industry coming to the area would be a brick factory. The site received (and continues to receive) the more hazardous of wastes, including heavy metals and PCBs. Interestingly, the current plan (as we go to press) of the City of Fort Myers, Florida, that has been submitted to the Florida Department of Environmental Protection provides that, after treatment at a Florida facility, the toxic sludge being removed from the Dunbar area of the City will be transported to the Emelle facility.

As discussed previously, the 1987 study of the United Church of Christ entitled "Toxic Waste and Race in the United States" reported results similar to those of the 1983 GAO study. The 1987 study, however, was nationwide in scope and included both hazardous waste facilities and "uncontrolled hazardous waste sites," the latter being an EPA classification that includes closed and abandoned disposal facilities, accidental spills, illegal discharges, dumps, and closed factories or warehouses where hazardous waste was produced, used, or stored (UCC 1987). This study found that "race was consistently a more prominent factor in the location of commercial hazardous waste facilities than any other factor examined" (UCC 1987, p. 15). For example, of the five largest hazardous waste facilities in the United States, three (Emelle, Alabama; Scotlandville, Louisiana; and Kettleman City, California) have minority (either African-American or Hispanic) populations in excess of 75%, and those three sites account for 40% of the total landfill capacity in the United States.

However, as Bullard has observed (1994c, p. 1041),

> Environmental justice is not just about facility siting. It also involves issues and concerns around pesticide exposure, lead poisoning, transboundary toxic waste dumping, shipping

risky technologies abroad, unequal protection, differential exposures, and unequal enforcement of environmental, public health, civil rights, and housing laws.

The Dumping of Waste and Abandonment of Waste Sites

As indicated by the two case studies, aside from a decision to place a toxic waste site in a particular location, toxic or hazardous waste is often dumped in rural areas inhabited, at least in the southeastern United States, by economically disadvantaged African-Americans.

Pollution

The third major type of action is the existence of facilities that produce air or water pollution near communities or predominantly minority or low-income populations, and the failure of regulators to curb intense pollution in locations inhabited by economically disadvantaged persons and racial minorities. In the southeast United States, the usual minority is African-Americans. In the southwest and western United States, typically, the affected minorities are Hispanics and Native Americans. In other geographic areas, it is "all of the above."

Environmental Justice and Its Relation to Environmental Crime

The failure to achieve environmental justice or environmental injustice, per se, is not a crime. However, conduct that is indicative of environmental injustice may, and often does, involve criminal conduct, but the act is criminal because it violates an environmental crime statute and not because it is environmental injustice. For example, the intentional spraying of toxic waste on a roadside without a permit is a criminal violation of the Toxic Substances Control Act (see 15 U.S.C. §2606).

Have We, or Can We, Achieve(d) Environmental Justice?

Executive Order 12,898 was issued in 1994. The EPA was, and is, the principal agency responsible for the implementation and enforcement of federal environmental policies, including environmental justice. In 2005, the Government Accounting Office determined that the EPA had not followed recommendations for the consideration of environmental justice in its adoption of three rules pursuant to the Clean Air Act; that "the agency's claim that its rules have resulted in better air quality nationally is irrelevant in this context [i.e., environmental justice]" (Trial, 2005, p 83).

The following year, the EPA Office of the Inspector General issued a report highly critical of the EPA for failing to implement environmental justice reviews.

In 2007, the General Accounting Office again criticized EPA rules weakening the Toxic Release Inventory program because it would reduce the amount of information used to assess environmental justice.

In 2011, 17 federal agencies signed a memorandum of understanding on environmental justice. Each of the agencies agreed to publish its environmental justice strategy on its webpage, as well as annual progress reports on the implementation of that strategy. The federal courts, however, have not recognized a private cause of action based on Executive Order 12,898 or agency regulations.

The furthest some of the courts have gone is to review an agency's consideration of environmental justice in its environmental impact review (see *Communities Against Runway Expansion, Inc. v Federal Aviation Agency*, 355 F.3d 678 (D.C. Cir. 2004).

In 2017, the Michigan Civil Rights Commission issued its report concerning the lead contamination crisis in the water in Flint, Michigan, concluding that the disparate treatment of the Flint crisis was the result of "systemic racism" (Michigan Civil Rights Commission, 2017).

The only sound conclusion, therefore, is that achieving environmental justice is elusive; there is much yet to be done. The EPA's record concerning environmental justice has been spotty at best. One wonders whether the recommendation back in the 1990s that the mandates of Executive Order 12,898 be legislated and that judicial remedies be added should be reconsidered.

Environmental Justice Case Studies

Warren County, North Carolina, Is Dumped on Twice in the 1980s: The Beginning of the Environmental Justice Movement

Warren County is situated on North Carolina's northerly border with Virginia west of I-95, northwest of Rocky Mount, N.C., and north, northeast of Raleigh, N.C. Though comprised of 444 square miles in territory, its 1980 population was slightly more than 16,000. The County's population was 60% African-American, compared to 22% for the State. Warren County ranked 97 out of 100 North Carolina counties in per capita income.

In the 1970s and 1980s, Ward Transformer Company, Inc., engaged in the purchasing, rebuilding, and resale of electrical voltage transformers in Raleigh, North Carolina. At one time, Ward Transformer was one of the largest transformer rebuilders in the country. The transformers purchased for rebuilding contained oil with large concentrations of polychlorinated biphenyls (PCBs), a liquid toxic chemical. (The Toxic Substances Control Act of 1976 banned the manufacture of PCB.) Over time, Ward Transformer stored a large volume of the PCB-infused oil and, following the implementation of EPA regulations governing the storage and disposal of toxic waste (including those with PCBs), the company sought to dispose of some 7,500 gallons of oil with concentrations of PCB. Ultimately, Robert Ward, the company's board chairman, contracted with Robert Burns, a long-time friend, to remove the oil to a storage facility in Youngsville, Pennsylvania. When Burns realized that his plans for the oil would not be profitable, he contacted Ward, who approved Burns' plan to dispose of the oil by dumping the oil on rural sites in North Carolina. At Ward's suggestion, Burns outfitted a truck with a large tank and spray nozzle at the Ward Transformer facility and proceeded to dump two loads of the oil on sandy soil at the impact range for Fort Bragg, located in the southern portion of North Carolina. The truck got stuck in the sandy soil, and the sand did not absorb the oil as well as expected. Burns then decided to spray the oil along rural roadsides in North Carolina, and with Ward's approval, redesigned the truck and spray nozzle so the nozzle could be operated by a passenger in the truck, and the vehicle proceeded along the roadways. During the month of July and until August 4, 1978, Burns' two sons would pick up the oil at Ward Transformer, select a rural section of road, and dump the oil in a 4- to 6-inch band along the road. The dumping consumed 240 roadside miles in 14 counties of North Carolina.

The band of oil alongside the road was soon detected by federal and State authorities. Burns confessed to the unlawful dumping, and Ward ultimately was convicted in federal district court of multiple counts of unlawful disposal of toxic substances and aiding and abetting the unlawful disposal of toxic substances. That conviction was affirmed on appeal (*United States v Ward*, 676 F.2d 94 [4th Cir. 1982]), and the U.S. Supreme Court denied certiorari (*Ward v United States*, 459 U.S. 835 [1982]).

But, what to do with the contaminated soil along the roadways? At first, the North Carolina Department of Transportation (NCDOT) covered the contaminated strip of soil with sand and issued advisories not to eat produce grown nearby. Later, the NCDOT covered the soil with a charcoal solution and asphalt to minimize the odor and to keep the soil in place. The NCDOT proposed to remove the tainted soil temporarily to its maintenance yard in Warrenton, N.C., the county seat of Warren County, but that was precluded by a court injunction. The State proposed several solutions to deal with the soil. Two of the alternatives were not approved by the EPA, two others were considered cost prohibitive, leaving only two other alternatives: in-state burial of the waste and trucking the waste to an out-of-state disposal site. The state opted to bury the waste at a single site within the state. Initially, some 90 potential sites were identified but most were eliminated because they would not satisfy EPA regulations. Ultimately, the review was reduced to two sites for an in-State landfill: land owned by the Pope family near the community of Afton in Shocco Township, Warren County, and a portion of the existing Chatham County sanitary landfill. By comparison, the population of Shocco Township was 66% (84% in the community of Afton) African-American; Chatham County was 27% African-American and 72.8% white. Thirty-two percent of the population of Shocco Township had family income below the poverty level, and of that percentage, 90% were African-American.

The Chatham County landfill satisfied EPA requirements for the disposal of the contaminated soil; the Warren County site owned by the Pope family did not. After Chatham County (located southwest of Raleigh-Durham, N.C.) residents voiced strong opposition, the Chatham County Board of Commissioners withdrew its offer to sell a portion of its landfill site. According to the N.C. Attorney General, the State did not have the power of an eminent domain. That left only the Pope site for consideration. The State convened a public hearing on January 4, 1979. Because the Pope site did not meet EPA requirements, the State sought, and the EPA ultimately approved, three variances: (1) elimination of the requirement of 50 feet between the landfill and groundwater; the Pope site was 7 feet between groundwater and the proposed landfill; elimination of the requirement of an artificial liner based on the view that soil compaction would be sufficient to prevent leakage; and elimination of an underliner leachate collection system. After this hearing, a group of concerned residents met with N.C. Governor James Hunt, who had not attended the public hearing. Hunt insisted that landfill construction would proceed safely. The same group of residents also met with EPA officials in Washington, D.C., and later initiated the creation of the Warren County Citizens Concerned with PCBs.

When it received notice that the EPA had approved the State's landfill permit application, Warren County initiated a lawsuit in federal district court seeking to prevent the use of the Pope land for PSB landfill purposes. Initially, the Court granted a preliminary injunction halting further activity. On November 25, 1981, however, the Court granted judgment dismissing the County's action (*Warren County v State of North Carolina*, 528 F.Supp. 276). That same day, the Court also dismissed an action seeking the same relief that was commenced by landowners whose property adjoined the proposed site (*Twitty v State of North Carolina*, 527 F.Supp. 778).

In mid-September, 1982, some 4 years after the dumping of PCB-laden oil ended, some 60,000 cubic yards of contaminated soil was collected and dumped in the 22-acre landfill. The remaining 120 acres surrounding the landfill area was designed to serve as a buffer. The process did not run smoothly. Protestors, including civil rights activists from across the nation, occupied the roadway, many actually lying down in the road only to be forcefully removed and arrested. More than 500 protestors, including a U.S. Congressman, were arrested in the month-and-a-half period during which some 7,000 trucks deposited contaminated soil at the landfill. The landfill was capped in late November 1982.

The "Cadillac of landfills" (as Governor James Hunt called it) developed problems. Bubbles were observed at the liner, and the cap had to be vented. A pump failed, and excess water accumulated at the landfill area. Testing performed in 1994 revealed low levels of dioxin, a byproduct of PCB, in monitoring wells up and downhill from the landfill. Ultimately, by the end of 2003 and at a cost of $18 million, the site was decontaminated to a level greater than federal standards.

The City of Fort Myers, Florida, Dumps on Dunbar in the 60s: The Cleanup Struggle Continues

On May 7, 1962, almost two decades before the environmental justice movement began, the City Council of the City of Fort Myers, Florida, passed a resolution authorizing the purchase of several parcels of land at a cost of $1,600 for a "municipal purpose," specifically the disposal of lime sludge from its water treatment plant. The land, then known as City View Park, comprised a 4-acre city block in the area known as Dunbar, a predominantly African-American community east of railroad tracks that divide predominantly white and black communities. The plan was to dump 25,000 cubic yards of the waste into pits dug deeper than the water table. Shortly after the acquisition, the City began to dump truckloads of lime sludge into the area and apparently continued to dump lime sludge for several years.

After the sludge dumping, the pits were covered with soil. The area was not fenced until June 2017, so during the 1970s and for generations, children played in the area. They described the soil as squishy orange slag, like soft clay. Although the City was aware of the arsenic contamination at some point in the late 1990s and the State Department of Environmental Protection (DEP) was aware in 2006, residents in the area were not informed of the contamination until June 12, 2017, when an article appeared in the local newspaper, *The News-Press*.

In 2007, the City explored the possibility of showcasing the site as a housing project of affordable homes called Home-a-rama. Arsenic levels above EPA acceptable limits were detected on the site, and the development never proceeded.

Over time the sludge seeped into the soil and apparently the aquifer. Residents complained, and the testing of tap water revealed the presence of arsenic. The City hired a private engineering firm to undertake soil sampling and to assess the environmental danger to homeowners in the Dunbar area.

It gets worse. In August 2017, *The News-Press* reported on the discovery of another site used for the deposit of toxic lime sludge. Unlike the Dunbar site, the Eastwood site is isolated from the population. The City, however, had designated that site, the Eastwood site, as a brownfield. The City did not make, and to date has not made, the same designation for the Dunbar site.

Although requested by Dunbar residents, the Florida State Department of Environmental Conservation, and Bill Nelson, a U.S. Senator representing Florida, the U.S. EPA has not provided assistance or enforcement support. There is no mention of the Fort Myers site in the EPA's 2017 Report (EPA Report, 2017) or in the EPA's 2020 Action Agenda (EJ 2020 Action Agenda).

In May 2018, the State DEP ordered the City of Fort Myers to excavate and remove the sludge; either transport it to an approved hazardous waste site or secure the sludge to prevent public contact, including dust emissions; and test soil samples at the sides and bottom of the sludge pits to ensure that all of the contaminated soil has been removed. The City has proposed to excavate the sludge by December 2018.

Finally, on September 4, 2018, the City of Fort Myers appropriated funds to remove all of the lime sludge from the site (News-Press, 2018). The removal process has been completed.

Discussion Questions

1) The concept of environmental justice developed into a movement in the early 1980s. There are many who contend that the progress of the movement has been slow. What are some of the reasons for the movement's inability to gain traction?
2) The case studies deal with environmental events that occurred two decades apart. Discuss the similarities and differences between the cases as they relate to environmental justice issues.
3) There is considerable debate whether the City of Flint, Michigan, water crisis was the result of environmental racism or injustice. Discuss factors that should be considered in making that determination.

References

Bryant, B. (1995). *Environment justice: Issues, policies, and solutions* (2nd ed.). Washington, DC: Island Press.
Bullard, R. (1994a). *Dumping on Dixie: Race, class, and environmental quality* (2nd ed.). Boulder, CO: Westview Press.
Bullard, R. (1994b). The legacy of American apartheid and environmental racism. *St. John's Journal of Legal Commentary, 9,* 445.
Bullard, R. (1994c). Environmental racism and invisible communities. *West Virginia Law Review, 96,* 1037. Washington, DC: Island Press.
Bullard, R. (1994d). Overcoming racism in environmental decisionmaking. *Environment, 36*(4), 10.
Carter, E. (n.d.). Internet encyclopedia of philosophy. *The American Environmental Justice Movement*. Retrieved October 8, 2018 from https://www.iep.utm.edu/enviro-j/
Cole, L. (2007). Environmental justice comes full circle: Warren County before and after. Golden Gate University Environmental Law Journal, 1, 9.
Communities Against Runway Expansion, Inc. v Federal Aviation Agency, 355 F.3d 678 (D.C. Cir. 2004).
Executive Order 12,898. (1994). 3 C.F.R. §859 (1995).
Frye, R. (1993). Environmental injustice: The failure of American civil rights and environmental law to provide equal protection from pollution. *Dickinson Journal of Environmental Law & Policy,* 3, 56
Heagerty, M. (2010). Crime and the environment—Explaining the boundaries of environmental justice. *Tulane Environmental Law Journal,* 23, 517.
Lynch, M., Burns, R., & Stretsky, P. (2014). *Environmental law, crime, and justice*. El Paso, TX: LFB Scholarly Publishing.

Michigan Civil Rights Commission. (2017). *The flint water crisis: Systemic racism through the lens of Flint.* Retrieved on May 30, 2019 from https://www.michigan.gov/documents/mdcr/VFlintCrisisRep-F-Edited3-13-17_554317_7.pdf

The News-Press. (2017, June 23). *City of Fort Myers dumped toxic sludge in Dunbar.* Retrieved on July 19, 2018 from https://www.news-press.com/story/news/2017/06/12/city-fort-myers-dumped-toxic-sludge-dunbar/369451001/

The News-Press. (2018, March 13). *DEP gives Fort Myers ultimatum on sludge.* Retrieved on July 19, 2018 from https://www.news-press.com/story/news/2018/03/12/breaking-dep-gives-fort-myers-ultimatum-toxic-sludge/417439002/

Trial Magazine. (2005, October). EPA rule-making gives environmental justice short shrift, GAO says. *Trial, 41,* 81.

Twitty v State of North Carolina, 527 F.Supp. 778 (E.D.N.C. 1981).

U.S. Commission on Civil Rights (Commission). (2003). *Not in my backyard: Executive order 12,898 and title VI as tools for achieving environmental justice.* Retrieved October 15, 2018 from https://www.usccr.gov/pubs/envjust/ej0104.pdf

United Church of Christ (UCC). (1987). *Toxic wastes and race in the United States.* Retrieved on October 9, 2018 from http://uccfiles.com/pdf/ToxicWastes&Race.pdf

United Church of Christ (UCC). (2007). *Toxic wastes and race at twenty, 1987–2007.* Retrieved on October 9, 2018 from http://d3n8a8pro7vhmx.cloudfront.net/unitedchurchofchrist/legacy_url/7987/toxic-wastes-and-race-at-twenty-1987-2007.pdf?1418432785

United States Environmental Protection Agency. (2016). U.S. EPA's environmental justice strategic plan for 2016–2020, EJ 2020 Action Agenda. Retrieved from https://www.epa.gov/sites/production/files/2016-05/documents/052216_ej_2020_strategic_plan_final_0.pdf

United States Environmental Protection Agency. (n.d.). EPA environmental justice FY 2016 progress report (EPA report). Retrieved from https://www.epa.gov/sites/production/files/2017-09/documents/fy16_ej_progress_report.pdf

United States General Accounting Office. (1983). *Siting of hazardous waste landfills and their correlation with racial and economic status of surrounding communities.* Washington, DC: Government Printing Office. Retrieved September 1, 2018, from http://archive.gao.gov/d48t13/121648.pdf.

United States v Ward, 676 F.2d 94 (4th Cir. 1982).

University of North Carolina Exchange Project. (2006). *Real people – Real.* Retrieved from http://wayback.archive-it.org/3491/20180622142705/http://www.exchangeproject.unc.edu/documents/pdf/real-people/Afton%20long%20story%2007-0426%20for%20web.pdf

Warren County v State of North Carolina, 528 F.Supp. 276 (E.D.N.C. 1981).

Chapter 9

Environmental Crime Around the World

In large part, the previous chapters of this book dealt with the problem of crimes against the environment as they relate to the United States. This chapter widens this perspective significantly by addressing environmental crime and its control throughout the rest of the globe. In it, we explore the dimensions of crimes against the environment in other nations as well as international environmental crime, crimes that recognize no official national borders.

Until now, this book has primarily concentrated on crimes involving the illegal disposal of substances having chemical compositions that are harmful to the environment. Certainly, these types of crimes are of great concern to international organizations like the United Nations Environmental Programme (UNEP) and Interpol, but these are not the only threats that face the international community. The United Nations and Interpol interpret environmental crime as illegal activities harming the environment and aim at benefiting individuals or groups from the criminal exploitation of the environment. This includes but is not limited to serious crimes and transnational organized crime. Illegal wildlife trade is near the top of the list, accounting for up to $23 billion per year. Raising the awareness of this problem has been the wholesale slaughter of elephants and rhinos. These crimes also include illegal activities in the forestry sector, the illegal exploitation and sale of gold and minerals, and illegal fisheries/fishing. On an international scale, trafficking in hazardous waste and chemicals is estimated to be valued at between $10 and $12 billion. Some of the wealth generated from the illegal exploitation of natural resources is used to support non-state armed groups and terrorism. When one considers the international scale of environmental crimes, one must also consider the impact on costs to future generations. The illegal disposal of chemicals, deforestation, and illegal fisheries can lead to a depletion of ecosystem services. This is manifested in the pollution of air and water, weather mitigation, and food depletions. The rise of transnational organized crime has witnessed an evolving shift from illegal activities like drug and human trafficking, arms trafficking, and the proliferation of counterfeit products to illegal environmental transgressions. These include trafficking in hazardous waste and chemicals, forest products, minerals, and illegally extracted gold. These activities also include illegal trafficking in giant clams and pangolins. Taken as a whole, the annual growth rate of these crimes is estimated as high as 21% to 28% (UNEP, 2016).

Illegal Wildlife Trade

Pangolins

Pangolins are commonly known as scaly anteaters. They come in eight different species, four of which are located in Africa and the other four in Asia. Two Asian species are listed as critically endangered, and another two are listed as endangered. The international illegal trade in the species is focused on the Asian species. Much of the illegal trade here stems from Southeast Asia. China and Vietnam drive the illegal trade for scales used for medicine and meat considered to be a luxury. It is estimated that around one million of these species have been taken from the wild in the last ten years. Myanmar is also a main source and transit country due to its geographic location and weak government. Seizures of pangolin parts in Zimbabwe reached a total of eight tons in 2015. Other large seizures in 2015 occurred in Uganda, the Republic of Congo, Kenya, and Nigeria (UNEP, 2016).

Elephants

African elephants numbered about 1.3 million in 1972. In 2013, that number was reduced to 473,000. Between 2009 and 2012, it was estimated that approximately 100,000 elephants were illegally poached in Africa. The poaching of these elephants generated an average of 210 metric tons of illegal ivory per year. Approximately 75% of elephant poaching occurs in central and South Africa, and 25% occurs in East Africa. A poaching hot spot is the Republic of Tanzania. In Asia, it is estimated that poaching accounts for about 40% of all elephant deaths. Male Asian elephants are particularly prized because they have larger tusks than African elephants (UNEP, 2016).

Rhinos

Poaching has been particularly acute for critically endangered African rhinos, both black and white. White rhinos are primarily concentrated in South Africa, and black rhinos are located in Namibia, Swaziland, and Zimbabwe. In 2015, poaching of rhinos in Africa reached its highest peak in decades, amounting to over 1,300 rhinos. It is estimated that this amount will be reached if not exceeded by the end of 2017. Over 50% of rhinos killed in South Africa emanate from Kruger National Park. In Asia, Nepal has been a hot spot for rhino poaching. Over the years, poachers have become more sophisticated. More of the poachers are using high-powered .375 and .458 rifles, crossbows, and helicopters. Highly professional poaching has been commonly referred to as "khaki colored crime" and is said to be carried out by industry insiders with veterinarians (UNEP, 2016).

Forestry Crimes

There is clear evidence that transnational organized crime is becoming increasingly involved in forest crimes. These crimes include illegal logging, the laundering of illegal tropical timber through "fraud" plantations, and the laundering of timber through paper mills and palm oil plantation front companies. In addition, transnational organized crime is responsible for a large proportion of the smuggling of rosewood. Estimations are that up to 85% of all illegal tropical wood entering the United States comes in the form of paper, pulp, or wood chips. Illegal actions here include obtaining false road transport and logging permits; faking documents certifying production from plantations; obtaining cross-border permits through bribes; mixing illegal timber with legal local timber; and processing roundwood into paper chips or paper before exporting, making traceability very difficult or impossible.

Illegally harvested rosewood from West Africa is a growing environmental crime internationally. Being a hardwood used to produce furniture, flooring, and musical instruments because of its special acoustic quality, rosewood is a highly sought commodity. The main destination for illegally harvested rosewood is China, where the Hongmu furniture industry thrives. Recently a group of West African countries coordinated with Interpol to address the illicit rosewood trade which has been depleting rosewood trees. Those countries include Togo, Senegal, Ghana, and Gambia (UNEP, 2016).

Fishing

Estimations are that an average of 15% to 18% of the global catch of fish is caught illegally each year, undermining food security, fishery management, and biodiversity. Fisheries are particularly susceptible to criminal exploitation because much of the fish is caught in areas outside of national jurisdictions and in remote areas that are outside the jurisdiction of local law enforcement agencies. Offenders are mobile and take advantage of vague international legal frameworks. It is common for private firms to register vessels in vulnerable flag states so that the vessel owners can register their vessels in countries unwilling to enforce the law. Crimes associated with this industry include criminal abuses of registries, customs regulations, tax frauds, and forced labor. In addition, offenses extend into the areas of illegal arms trafficking and terrorism. Australian and French navies, at the beginning of 2016, detected dhows containing weapons en route to Somalia. Illegal explosives used for blast fishing have also been connected to terrorist activities in Tanzania (UNEP, 2016).

Waste Pollution

Global illegal waste trafficking remains a critical problem throughout the world. It is estimated that over 400 billion tons of hazardous waste are trafficked each year. Much of that hazardous waste trafficked throughout the world is done so through organized crime groups. The primary entities for tracking these wastes are the Basel, Rotterdam, and Stockholm Conventions. In addition, the United Nations' *Solving the E-waste Problem (StEP) Initiative* on electronic waste (E-waste) is responsible for tracking electronic waste internationally. To bypass the law enforcement authorities, these wastes are often classified as other items. The mechanism used to complete these illegal acts is the use of "non-hazardous" waste codes for hazardous wastes and by employing the use of product codes for hazardous wastes. Africa and Asia are primary destinations for large shipments of electronic hazardous wastes. Due to the fact that countries in Africa and Asia lack resources for effective control, these countries are often exploited by organized crime. Much of Western Europe is the area of origin of global illegal waste trafficking. Other areas where illegal waste exportation has been proven include Russia, Lithuania, Ukraine, Croatia, and Albania. Two main trafficking destinations have found to be Ghana and Nigeria. The Italian national anti-Mafia directorate has cited many Italian cases of the illegal disposal of hazardous waste that involved fraudulent authorizations, frequently issued by corrupt government officials. Europol reported that illegal waste trafficking is frequently facilitated through cooperation with legitimate businesses. Those businesses include financial services and import/export operations in metal recycling sectors. This trafficking often depends upon the use of specialists in document forgery for the acquisition of permits (Europol, 2011).

The following subsections offer greater details on special waste pollution problems facing five diverse parts of the world: Italy, Latin America, Russia, Ukraine, and East/South-East Asia.

Italy

Historically, illegal waste trafficking and disposal have been a problem in Italy. In March of 2016, Italian police arrested five Eni Oil staff in a waste trafficking case. Eni is an Italian multinational oil and gas company headquartered in Rome. Five staff at a treatment plant operated by Eni were charged with illegal waste trafficking. The arrests were the result of a two-year-long investigation by regional anti-mafia prosecutors into waste management at the plant in southern Italy located in the Val d'Agri oil concession area in the Basilicata region. Production at the field of around 75,000 barrels per day was promptly halted on the heels of the arrests. In 2015, waste trafficking rose to national attention in Italy with the report that at least 14 individuals had been carrying out a major illegal waste trafficking operation in Rome, Naples, and Salerno. The Anti-mafia District Directorate (DDA) in Rome targeted alleged members of the Camorra, the crime syndicate based in the southern region of Campania. Prosecutors contended that the criminal group was led by Camorra boss Pietro Cozzolino. Cozzolini's operation was said to be prominent in the deprived areas of Ercolano and Portici in the province of Naples. The wastes illegally trafficked were described by prosecutors as including large quantities of textile wastes and hazardous wastes emanating from other countries (Jewekes, 2016).

Hazardous waste crime in Italy has been massive. The Italian environmental group Legambiente has reported that more than 29,000 environmental crimes were committed in Italy in 2013. And, in 2013, the illicit trafficking of toxic waste alone was said to be worth close to $4 billion.

Legambiente has reported that at least 40% of environmental crimes have been committed in the southern regions of Campania, Puglia, and Calabria, followed by the northern Lombardy region. The report cited the illicit disposal of garbage as being intense in Campania, where Camorra clans have been identified as the culprits of dumping and burning toxic waste for approximately the last 20 years. The toxic dumping resulted in an environmental disaster in the region, with higher rates of deaths from cancer and congenital malformations in newborn children registered among the population of the most affected areas, especially between Naples and the city of Caserta, an area now referred to as the "Triangle of Death." In this area, women were found to contract breast cancer unusually early, men had developed high rates of lung cancer despite never having smoked, and children were being born with Down's Syndrome in comparatively young mothers. In an investigative journalism piece by Shire Cohen, local citizens described how the mafia had dumped extraordinary volumes of contaminated industrial waste and later obtained backdated permission for their actions. Hazardous materials were abandoned on prime agricultural land, next to a car dealership, with bingo halls and furniture stores nearby, and just a few hundred yards from a town of 40,000. By 2016, thousands of similar dumps had been discovered in canals, caves, quarries, wells, under fields, and beneath overpasses. "Midnight dumping" by truckers became common, as did the widespread burning of the wastes. Trucks would turn up at night, waste would be emptied, and then huge fires started. In its own backyard, the Camorra mafia was responsible for burying barrels of wastes, driving containers filled with wastes into rivers, and spreading chemical sludge on fields as "fertilizer" (Birrell, 2016). The present problem began as a result of a 1980 earthquake that took the lives of close to 3,000 people and rendered over 250,000 homeless. The monetary aid that followed wound up being controlled by organized crime aid as the construction industry was ruled by the mafia.. Businessmen with organized crime connections who owned waste dumps realized big profits could be made hiding industrial waste in debris (Birrell, 2016; Yardley, 2014).

For a long time, major mafia organizations in Italy have been expanding their illegal businesses to dumping and trafficking of hazardous wastes. Trafficking in toxic waste has developed in Italy since the second half of the 1980s. Initially, it was undetected and almost invisible to Italian society at large until the second half of the 1990s. Progressively this highly lucrative market expanded quickly due to the growth in the special hazardous waste production and high profits and low risks. At the start of the 1990s, the first major trafficking route was detected to be controlled by a Camorra criminal group involved in many diverse criminal enterprises.

Illegally handled waste is produced by several sectors like the steel industry, chemical industry, pharmaceutical companies, and hospitals. The protocol for the treatment and disposal of hazardous waste mirrors the system in the United States to an extent, in that wastes must be properly transported, treated, and disposed following regulatory laws. If the waste generators do not have the capability of treating the wastes itself, it must contract with an approved entity to treat and dispose of the waste properly within the law (Massari and Monzini, 2004).

The "mafia-type" organizations that are responsible for illegal hazardous waste disposal and trafficking in Italy are also found to be involved in the legitimate handling of those wastes as businessmen. The mafia-type clients are involved in illegally handling hazardous waste throughout the entire nation. These individuals have demonstrated well-established business relationships with business entrepreneurs in northern and central regions of Italy. Criminal investigations in Italy have uncovered notable conspiracies between waste producers and collection and transport companies, and significant outsiders like laboratories. Like cases explained earlier in this book in states like Maryland and New York, farmers in Italy have been known to provide areas of their open space for makeshift dumping sites. Investigations in Italy have also exposed sophisticated systems for falsifying transport and processing documents, and bribery of public officials. The structure of trafficking in waste is quite simple, with a maximum of three or four people at the core of the organization who access the manpower for diverse tasks and a large network of connections with various professionals and firms. The profile of the offenders matches that of criminal treatment/storage/disposal (TSD) facility operators of the past in the United States. As put by Monica Massari and Paola Monzini in their study of waste trafficking in Italy:

> few of them had previous criminal records, especially for activities related to environmental crime. In most cases they were simply entrepreneurs (working for waste producing companies, commercial intermediaries, shipping firms or final disposal sites) who came together to conduct illegal business related to waste disposal. Involvement in this market requires access to at least some know-how and contacts (such as specialized hauling companies, storage sites, recycling centers, treatment equipment, ad hoc dumping sites, cooperative laboratories and

so on), in order to circumvent the official procedures required by regulations. Individuals who traffic in waste materials are usually not part of the classic "underworld." (Massari and Monzini, 2004, p. 298)

Much like the 1980s environmental crimes depicted in the United States in Rebovich's *Dangerous Ground*, Italy's environmental crimes of the 21st century reflect layers of a variety of offender types. Besides the entrepreneurial white collar crime types, Italian environmental crime also encompasses "significant outsiders" such as truck drivers and other unskilled and skilled workers. The authors do not just single out waste transporters, but also identify secondary figures like "sentries" and "unloaders" who are critical in developing large businesses. Like TSD operators in *Dangerous Ground*, the upper level of the market in Italy is restricted to a number of professional traffickers. Middlemen are active in contacting special waste producers, locating the right businesses, and sealing important connections. What is sold is the ability to handle special waste and to do it at the lowest prices. The criminal skills reside in identifying the best potential clients and persuading them to purchase clandestine, but reliable, services. Central to the operations are agreements facilitated by professional intermediaries between waste producers, waste collectors, transporters, and representatives of waste management corporations. Massari and Monzini also found that other significant outsiders include chemists and counterfeiters (Massari and Monzini, 2004).

In the end, Massari and Monzini's research has revealed that Italian trafficking in toxic waste is diverse in that it is committed by a wide range of actors from the traditional mafia-type organizations to loose networks of individuals with no criminal backgrounds, belonging to different economic sectors. The outcome of the research underscores the pressures of supply and demand. Massari and Monzini's research uncovers a picture of a combination of corporate entities striving to save money illegally that are part of unlawful operations supplying services to environmental businesses, and partnerships between organized crime affiliates and government officials (Massari and Monzini, 2004).

Environmental crime in Italy orchestrated by organized crime groups has continued unabated into the second decade of the 21st century. This entails not just the dumping of toxic waste but also the trafficking of endangered species. Yearly, tens of thousands of environmental crimes are detected. The areas most affected have historically been Campania, Calabria, Sicily, and Puglia, the base of Italy's four distinct mafia organizations. The Tuscany and Sardinia areas have also been the scenes of criminal dumping. Law enforcement has singled out the complicity of corrupt local authorities to the environmental crime success of mafia dons. An area hard hit by environmental crime is Campania, the region around Naples (Squires, 2012).

More recently, Aunshul Rege and Anita Lavorgna have produced a study on a relatively new area of organized environmental crime in Italy. This is the illegal supplying of soil and sand. Soil, particularly sand, is a key ingredient for the manufacturing of concrete, brick, glass, and also computers, mobile phones, and credit cards. Consequently, soil and sand are highly desired. Unfortunately, existing regulated supplies are not sufficient enough to meet the great demand. This demand has prompted the illegal widespread mining of soil and sand. In Italy, the problems resulting from excessive and illicit soil and sand mining include the destruction of natural beaches, increased shoreline erosion, and economic losses due to a loss of coastal aesthetics affecting the tourism industry. Rege and Lavorgna tell us that in Italy, organized crime groups have become illegal suppliers of environmental resources harvested from within Italy. *Ecomafias*, as they are now called, actively engage in illegal soil mining primarily to meet the demands of the construction industry. This activity is closely related to illegal waste disposal in that wastes are subsequently illegally dumped into them, converting them into makeshift illegal landfills. Wastes dumped into these holes have included industrial wastes, tires, demolition material, and even nuclear waste (Rege and Lavorgna, 2016).

Latin America

A careful analysis of environmental challenges facing countries in Latin America demonstrates that the challenges emanate from a range of sources. Climate change has been identified as the reason Argentina has experienced severe rainstorms and extreme heat, resulting in a decline in the population of Magellanic penguins on Argentina's Punta Tombo peninsula. Likewise, rising sea temperatures have been reported as the reason there is an increase in the bleaching of coral reefs off of Belize. In Chile, air pollution and pollution from mining operations have added to problems of deforestation, soil erosion, and water shortages, endangering 16 mammal species and 18 bird species. Columbia has experienced crude oil spillage combined with the effects of the illegal production of coca. There is some evidence that there is a correlation between increased cocaine production in Colombia and the loss of rainforest ground due to the fact that jungles close to coca plots have a greater likelihood to be cut down. El Salvador is especially hard hit by environmental threats. These threats range from water contaminated with feces and the consequences of the exploitation of precious metals. The presence of over 30 major mining projects in El Salvador has led to toxic chemicals being wantonly dumped into water sources. The once crystal blue waters of Guatemala's Lake Atitlan now have a thick brown sludge layering the top due to waste dumping, particularly from agrochemicals. In Puerto Rico, an island that has limited disposal space, large amounts of solid and toxic wastes accumulate. In the wetlands of Puerto Rico, there is so much solid waste that it blocks the entrance of the Martin Pena Channel and can cut off the ocean's flow. And in Venezuela, urban oil pollution has devastated Maracaibo Lake. At one point, several oil leaks filled the lake with crude oil. Despite the fact that these leaks do not compare to the Gulf of Mexico BP tragedy, the accumulation of these leaks with other smaller leaks taking place over the decades seriously threatens the ecosystem in the lake (Baral, 2016).

A special environmentally acute problem has existed in Bolivia for years due to the rise of the gold mining industry. The global economic crisis of 2008 had a direct influence on gold mining in countries of the Amazon. Gold prices rocketed, leading to a rise in global demand and, with it, a dramatic increase in gold mining in the Bolivian Amazon. The main environmental culprit in pollution related to gold mining is mercury since it binds to metal and forms a heavy amalgam to allow the collection of small gold particles in water. A highly toxic compound called methylmercury is formed when mercury used in the process of mining spills into the water. This compound can be absorbed by fish, and when the fish are consumed by humans, methylmercury seeps into the bloodstream and quickly travels to other parts of the body where 95% of the substance is absorbed. Studies have shown that if this substance passes into the bloodstream of a pregnant woman it will ultimately be transported into the blood of the developing child, spread to the brain and other tissues, often resulting in damage to the fetus's nervous system. The United Nations Environment Program has reported that 55 tons of mercury were spilled into Bolivian water and soil in the year 2012. UNEP also estimated that Bolivia is responsible for 6% of all the mercury that is released into the environment as a result of gold mining (Berton, 2016).

Brazil is also a Latin American country that has felt the damaging consequences of irresponsible practices connected with mining operations. In 2016, Brazil's federal police filed formal charges against three mining firms: Samarco, Vale, and BHP Billiton. Along with seven executives, the firms were initially charged with committing environmental crimes related to a major dam collapse and resultant mine waste spill. Considered by many to be the country's worst environmental disaster in its history, the government of Brazil filed suit against the company for over $5 billion for compensation to victims who suffered damage to be used for cleanup operations. Samarco was responsible for operating the Fundao tailings dam, a cooperative project between Vale and the world's largest mining firm, the Australian company BHP Billiton. The companies were alleged to have been negligent in allowing the dam's waste reservoir to erupt, releasing a 530-mile toxic mudflow that traveled in waterways spanning two Brazilian states and a large part of the Atlantic coastline. Over 60 million metric tons of iron mining water and toxic sludge were released. Nineteen people died as a result of this tragedy. Brazilian federal prosecutors later filed homicide charges against 21 mining executives in connection with the catastrophe (HNGN, 2016).

Russia

One of the most imposing environmental threats to Russia is deforestation. This has emerged as such a serious problem because of extensive illegal logging in accessible woodland regions. Rates of illegal logging in northwest Russia and in Russia's Far East are at extraordinarily high levels leading to dangerous levels of erosion and carbon dioxide release. The illegal logging activities also negatively threaten a wide range of species in the boreal forests. Nuclear contamination of the countryside resulting from Russia's historical nuclear weapons program and nuclear energy sector is also a major concern. A large portion of Russia's first-generation nuclear reactors are close to reaching the end of their life span, posing increased risks for disastrous accidents as the reactors are continually used. Also, the country's nuclear weapons program has resulted in permanent damage to the environment, evidenced in southern Siberia and in parts of the Ural Mountains. The effects of serious environmental violations during the Soviet Union era are still seen in the continued degradation in the quality of Russian land and water. This is especially so in the industrial belt along the southern section of the Ural Mountains. On another front, the operation of hydroelectric dams on the Volga River has contributed to a significant decrease in the Volga's water volume, triggering a process in which pollutants are retained at a rate much higher than normal. It goes without saying that the present environmental policies of the Russian government have significantly improved since the days of the Soviet Union. But, Russia has been criticized for not being aggressive enough in enforcing its environmental laws. Some critics have noted that Russia has been active in dismantling their own environmental agencies. An example of this was the elimination of the State Committee for Ecological Matters, Goskomekologiya, and the transferal of its responsibilities to an agency restricted to operations developing natural resources (Smith, 2015).

Ukraine

The Ukraine conflict sits as the primary cause for deep, ongoing environmental problems in that nation. This is particularly so in the Donbas region, a region in eastern Ukraine that represents the most densely populated of all the country's regions except for the region where the city of

Kiev is located. The Donbas region was the site of the greatest amount of unrest during the 2014 Ukrainian revolution and Russian military intervention. Prior to 2014, this region had already been one of the most polluted areas in Ukraine because of the presence of coal mining and heavy engineering throughout the decades. In 2002, this region was home to over 300,000 tons of industrial waste per square kilometer. Environmental challenges as a result of the 2014 unrest include military damage to environmentally hazardous industrial sites that include industrial installations and mines. Between 10 and 20 mines were flooded as a result of the 2014 conflict, leading to severe water and soil contamination. The local infrastructure was also negatively affected by the conflict, resulting in improper waste disposal and limitations on proper pollution control services. The area has witnessed a startling increase in the number of fires due to ammunition explosions and the contributing factor of a buildup of dead vegetation in forests. The collapse of environmental governance in the conflict-affected areas has only compounded the environmental problems in this part of Ukraine. The resulting lack of governmental oversight has led to an increase in illegal logging and the looting of timber stocks by armed groups. In some cases, criminal networks have taken over and expanded coal mining operations. A great concern of the government is that lax controls over the eastern border produce a fertile ground for an increase in the illegal trafficking of hazardous wastes (Weir, 2015).

The western part of Ukraine has experienced environmental threats of another kind. The area surrounding the city of Kalush in the foothills of the Carpathian Mountains was the site of these threats. In 2013, an Israeli company named S.I. Group Consort won a contract to clean up a hazardous waste sludge pit at a cost of 1 billion Ukrainian hryvnia, roughly equating to US$125 million. The wastes were a byproduct of the region's potash mining industry that had accumulated for over 25 years. Backed by Ukraine's Ministry of Ecology and Naturalist Resources, the firm declared the job a success. However, in 2014, independent tests done by the Organized Crime and Corruption Reporting Project (OCCRP) discovered levels of hexachlorobenzene (HCB) contamination that were several thousand times the legal limit. Based upon public records, it was unclear exactly what was done at the site. It was revealed, eventually, that some of the waste was shipped to England and Poland. The waste that went to Poland was shipped to an incinerator in Gdansk that was ill-equipped to process 1,000 tons of waste annually. Consequently, thousands of tons of poisonous waste were stored improperly at the incinerator site. Meanwhile, back in Kalush, many residents still suffer respiratory problems. HCB typically enters the body through air or water contact and eventually spreads to all tissues. The chemical can remain in the body for many years. Environmentalists have predicted that chemicals from the wastes will ultimately seep into the groundwater of the Sivka River and from there into the Black Sea (Velychko, 2015).

East/South-East Asia

Electronic waste (E-waste) looms as one of the greatest environmental threats to Asia. In East and South-East Asia, the volume of discarded electronics rose almost two-thirds between the years 2010 and 2015. E-waste generation is growing fast in both total volume and per capita measures in these regions. This generation is driven by rising incomes and the increasing demand for new gadgets and appliances. The average increase in E-waste across all 12 countries and areas analyzed (e.g., Cambodia, China, Hong Kong, Indonesia, Japan, Malaysia, the Philippines, Singapore, South Korea, Taiwan, Thailand, and Vietnam) was 63% in the 2010 to 2015 period and totaled over

12 million metric tons. It has been reported by environmental experts that many of these countries lack adequate infrastructures to ensure proper environmental protection, which, in turn, raises the likelihood of improper recycling and disposal (UNU, 2017; Honda, Khetriwal, and Kuehr, 2016).

Pertinent information on the E-waste threat in East and South-East Asia is supplied through the report *Regional E-waste Monitor: East and Southeast Asia*, compiled by the United Nations University (UNU). The research shows rising E-waste quantities overtaking population growth in East and Southeast Asia and warns of improper and illegal E-waste dumping on the rise in most countries in the study regardless of the enactment of national E-waste legislation. Trends responsible for increasing volumes of E-waste in these countries can be boiled down to four main categories. First, technological innovations are responsible for the rapid introduction of new products, especially in electronics. These products include tablets, smartphones, and wearables like smart watches. Second, the increase in the number of consumers in industrialized countries in East and South-East Asia, coupled with the ongoing expansion of middle-class individuals who are able to afford more gadgets. The third is the decrease in the usage time of gadgets, resulting in older products becoming obsolete or incompatible at a quicker pace. And fourth is the increase in the importation of electrical and electronic equipment and subsequent increases in E-waste as they reach the end of their life span. The report also notes the growing presence of consumers, dismantlers, and recyclers who engage in illegal dumping, especially "open dumping," in which non-functional parts and residues from dismantling and treatment operations are released into the environment (UNU, 2017; Honda, Khetriwal, and Kuehr, 2016).

The report details the special danger posed by "backyard" recycling, particularly for the developing countries in the region. This is manifested in the mammoth business of conducting unlicensed and frequently illegal recycling practices literally taking place in backyards. The processes have proven to be dangerous to both the recyclers and the surrounding environment and are often patently ineffective in that they are usually unable to successfully extract the complete value of the processed products. In general, gold, silver, palladium, and copper are recovered from printed circuit boards and wires using "acid baths," hazardous wet chemical leaching processes. In addition, these "recyclers" employ the use of solvents such as sulphuric acid and use the leachate solutions for separation and purification processes. This concentrates the valuable metals and separate impurities, resulting in the release of toxic fumes. Even indirect exposure to the hazardous substances endangers the health of families of informal recyclers who live in the vicinity of the recycling processes as well as the health of those in nearby communities. Some research has demonstrated connections between exposure from improper treatment of E-waste and altered thyroid functions, reduced lung functions, negative birth outcomes, reduced childhood growth, negative mental health outcomes, and impaired cognitive development. The prognosis is that, short of the implementation of effective enforcement and control mechanisms, East and South-East Asia will experience continued E-waste threats for years to come (UNU, 2017; Honda, Khetriwal, and Kuehr, 2016).

Discussion

The central aim of Chapter 9 was to introduce the reader to the characteristics of environmental offenses that are taking place in different parts of the world. The early part of the chapter depicts a pronounced environmental concern that up to this point has not been fully discussed in this book. This environmental concern is the illegal wildlife trade. Illegal wildlife trade varies according to the wildlife indigenous to distinct global locations. Such illegal trade includes endangered species of pangolins in the Far East and elephants and rhinos in parts of Africa. In addition, food security and biodiversity are undermined throughout the world by illegal fishing that can account for over 15% of the global catch of fish. But the bulk of the chapter returns to an examination of environmental waste pollution and how it differs depending on the geographic region considered.

Illegal waste exportation has been rife in places like Lithuania, Croatia, and Albania. In some cases, the success of this criminal trafficking is dependent upon criminals who specialize in forging documents and permits that are meant to disguise criminal trafficking activities. It is not uncommon for organized crime to play a part in this criminal phenomenon. The chapter draws attention to areas that have been badly victimized by waste pollution. A case in point is Italy, where, particularly in southern regions, organized crime has been responsible for dumping and burning hazardous waste for decades. Present-day hazardous waste dumping in Italy has been found to mirror many of the characteristics previously explained as being part of the environmental crime culture in the northeastern United States during the 1980s.

In Latin America, air pollution and pollution from mining operation pollution is presenting an ominous danger to mammal and bird life. In some Latin American regions, the results of the spilling of crude oil have been worsened by the effects of the illegal production of coca. The improper disposal of mining wastes, especially as a consequence of the surge in gold mining, has been rampant in Latin American countries like Bolivia. In Russia, a proliferation of illegal logging has been responsible for the escalating levels of carbon dioxide release. And, in Russia, the environmental violations in the Soviet Union can still be seen today through the continued degradation in the quality of land and water in many parts of that nation. This is taking place in a land where the political climate is such that environmental protection has been deprioritized. In Ukraine, environmental challenges exist as the result of the political and military unrest in 2014 as well as new challenges presented by the improper disposal and cleanup at Ukrainian mines. Finally, the chapter illustrates the special problem facing East and South-East Asia in the form of the improper/illegal handling of E-waste.

The United Nations Environmental Programme has expressed that while the definition of "environmental crime" is not universally agreed upon, it is generally considered a collective term to describe unlawful actions that damage the environment and are in some way designed to benefit individuals or groups for the exploitation of, damage to, or trade, or theft of natural resources. This definition aptly fits the activities described in this chapter that are now occurring throughout the world. UNEP points out that these global activities are increasingly posing a dangerous threat to the foundation of sustainable development and security. The organization emphasizes that environmental crime is expanding exponentially and will increasingly endanger wildlife populations along with entire ecosystems. The organization also strongly advocates a concentrated action and information sharing effort to effectively thwart the rise in environmental crimes across the globe (UNEP, 2016).

Discussion Questions

1) This chapter explored the characteristics of environmental offenses in different parts of the world. While some countries experienced types of criminal activities that are distinct, there were some that can be considered "cross-cutting" offenses in that the same type of criminal activities were found to occur in very different geographical regions. Identify one of these cross-cutting offenses, the regions in which they occur, and explain key characteristics of the

offenses and how they are committed. What actions do you believe government oversight can take to rein in these offenses? Do international aspects of the offenses hinder effective environmental control? Why or why not?

2) Environmental violations in Italy were found in some cases to be facilitated through organized crime groups. How might the involvement of organized crime in the commission of environmental violations present special problems for law enforcement? How would you compare the characteristics of environmental crime in Italy in the 21st century to environmental crime in the United States in the 20th century?

Case Study: Environmental Problems in Tunisia – "A Right to Breathe"

Over time, the country of Tunisia has been beset by environmental problems from many directions. This North African country bordering the Mediterranean Sea and the Sahara Desert is the northernmost country in Africa. In terms of area, it covers 64,000 square miles, and it is home to over 11 million people. What once was a beautiful nation has now become seriously damaged by drought, climate change, and illegal dumping. A popular tourist location, the island of Djerba, is now pockmarked with garbage dumps. A noted environmental activist in Tunisia, Arifet Mohammed, has reported that, at present, there seems to be very little control at all of the environmental threats to the country. As he tells it, rare birds, like the el-hbara and el-ghzel, are widely hunted. In addition, he asserts that private companies have taken advantage of Tunisia's impotent central authority in a failing economy by bringing in hunters who illegally poach gazelles and birds for profit. Presently, there are a number of lawless zones in Tunisia that are referred to as "black spots." Poachers are prevalent in the southern part of Tunisia, where they can hunt endangered species that are protected under international law according to the Bonn Convention of Migratory Species. One of the most targeted is the houbara bustard that is prized as a means to treat male impotency. Some of these hunters come from Saudi Arabia and set up hunting camps outside of Tataouine, not fearing the under-funded and unarmed government agents responsible for controlling and preventing poaching (McNeil and Addala, 2015).

Tunisia's Environmental Minister, Mohamed Salmane, has announced that the country is too weak to protect many of its natural resources as a result of the Jasmine Revolution of 2010. The Environmental Minister has noted that the country continues to struggle with the effect of processing Tunisia's 2.2 million metric tons of annual waste. The weakening of the government in the post-revolution phases has resulted in severe environmental degradation. Massive pollution has resulted from the construction of a gypsum factory in Oued El Ghar. There were reports that a worker in the factory had been fired at the waste treatment plant after he secretly filmed environmental crimes being committed there and sent them to the media. It is alleged that the company had caused the deaths of local trees by omitting non-filtered exhaust and had been dumping waste treatment byproducts into pits. Environmental activist Arifet Mohammed has claimed that the company became indifferent to what and where they dumped after the revolution, resulting in near extinction of local agriculture (McNeil and Addala, 2015).

On a shallow area of the Mediterranean Sea (the Gulf of Gabes), construction is taking place on a low-lying archipelago there, Kerkennah, where the highest point is only 30 feet above sea level. Land is being lost in this area due to a rising water level of two feet a year. But that situation has not stopped construction. Marine biologists in Tunisia have stated that building codes for proper construction near the rising sea are ignored. Environmentally, this area can be particularly problematic during the winter when Mediterranean lunar tides flood much of the area. The result is that the land becomes flushed with salt that is lethal for much of the ecosystem. Like many coastal islands across the Mediterranean, this area is threatened by the rising sea, which is expected to flood this area totally by 2050. The Kerkennah archipelago will eventually be divided into islets. Industrial fishing has been blamed for the decimation of fish stocks here. Fishing and shipping companies continue to ignore maritime regulations, and builders continue to ignore general environmental regulations. Commercial bottom trawling methods used by multinational fishing companies in this area of the Mediterranean continue to present environmental problems and has also severed Tunisia's submarine connection to the World Wide Web. Such bottom trawling has become common around the world, threatening fish populations everywhere. The industrial overfishing has been a primary reason why a black market has been created that is skilled in exploiting loopholes in environmental regulations to feed markets around the globe (McNeil and Addala, 2015).

Kerkennah's sister island across the Gulf of Gabes, Djerba, is also experiencing pronounced environmental problems. It is heavily dependent on the mainland for food, water, and electricity. Drought here has become the primary reason for the elevated costs of water, animal feed, and food. While the island had once been an agricultural center and is currently still a vibrant tourism destination, it is suffering from lack of rain, construction growth of hotels, erosion from rising seas, and, finally, catastrophic disposal of untreated waste and sewerage in illegal garbage dumps. The proper disposal of wastes on the island comes at a high price. Currently, waste management companies are buying land from Djerban farmers, who are still hurting from the post-revolution economy, to turn the farms into unregulated and illegal dump sites. On top of this, climate change has had a serious impact on Djerba. Rain rarely comes, and increasingly the seasons have disappeared, replaced by summerlike temperatures throughout the year (McNeil and Addala, 2015).

Tunisia's city of Gabes is another, even more serious, matter. A company, The Tunisian Chemical Group, began constructing a phosphate refining complex north of the city for the purpose of exporting fertilizers and preservatives. This started a continual process of trains delivering raw phosphate from nearby towns to be loaded onto ships primarily bound for Europe. Business consistently improved, and Tunisia eventually took its place as the fifth-largest phosphate exporter in the world. During a time of mining instability, strikes hurt the industry that resulted, ultimately, in the hiring of more miners and increased production. In addition, after the Jasmine Revolution, the seaside refinery in Gabes drastically increased production, seriously damaging local communities and ecosystems. The *phosphogypsum* waste from the wet-acid refining process has become a grave problem. The improperly disposed wastes proved to be radioactive, containing uranium and radium. The Gulf of Gabes became the home of fish with deformed spines, and the effects of the wastes led to a lowering of passerine birds' breeding performances. Radiation levels in the area became dangerous. The beach at Gabes took on the moniker of *Chott al-Mout*, "The Death Shore." It has been reported that an overwhelming stench is persistently present there with the surf turning ever-changing shades of a foamy brown color, a dark slimy "muck" permeating throughout. Public protests eventually began to sprout. At one protest, one man carried x-rays and medical bill copies, underscoring his health-related difficulties as a result of the pollution. And another strikingly lamented the loss of their former "paradise" and carried a sign that said in English, "We have the right to breathe in a healthy environment" (McNeil, 2013).

References

Baral, S. (2016, April 22). Earth Day 2016: Find out what environmental problems 20 Latin American countries face. *Latin Times*. Retrieved from http://www.latintimes.com/earth-day-2016-find-out-what-environmental-problems-20-latin-american-countries-face-163309

Berton, E. (2016, October 28). Gold rush in Bolivia sparks conflict between minors in the community. *Mongabay*. Retrieved from https://news.mongabay.com/2016/10/gold-rush-in-bolivia-sparks-conflict-between-miners-and-the-community/

Birrell, I. (2016, June 24). Mafia, toxic waste into a deadly cover-up in an Italian paradise: They have poisoned our land stolen our children. *The Telegraph*. Retrieved from http://www.telegraph.co.uk/news/0/mafia-toxic-waste-and-a-deadly-cover-up-in-an-italian-paradise-t/

HNGN (2016, January 15). Brazil charges mining companies with environmental crimes. *Headline and Global News*. Retrieved from http://www.hngn.com/articles/169852/20160115/brazil-charges-mining-companies-environmental-crimes.htm

Honda, S., Khetriwal, D., & Kuehr, R. (2016). Regional E-waste monitor: East and Southeast United Nations University & Japanese Ministry of the Environment. Retrieved from http://ewastemonitor.info/pdf/Regional-E-Waste-Monitor.pdf

Jewekes, S. (2016, March 31). Italian police arrest five Eni staff in waste trafficking inquiry. *Reuters*. Retrieved from http://www.reuters.com/article/us-eni-arrests-idUSKCN0WX2J6. Reuters

Massari, M., & Monzini, P. (2004). Dirty business in Italy: A case study of illegal trafficking in hazardous waste. *Global Crime, 6*, 3, August-November.

McNeil, S. (2013, June 14). Pollution in Gabes, Tunisia's shore of death. Al Jazeera. Retrieved from http://www.aljazeera.com/indepth/features/2013/06/20136913247297963.html

McNeil, S., & Addala, R. (2015, December 29). Environmental crimes run rampant. Al Jazeera. Retrieved from http://www.aljazeera.com/indepth/features/2013/11/environmental-crimes-run-rampant-tunisia-201311251180934679.html

Rege, A., & Lavorgna, A. (2016, May 19). Organization, operations, and success of environmental organized crime in Italy and India: A comparative analysis. *European Journal of Criminology*, 14, 2.

Smith, B. (2015, July 18). Russia: Environmental issues, policies and clean technology. *Azocleantech*. Retrieved from http://www.azocleantech.com/article.aspx?ArticleID=542

Squires, N. (2012, July 9). Mafia making billions from environmental destruction. *The Telegraph*. Retrieved from http://www.telegraph.co.uk/news/worldnews/europe/italy/9386961/Mafia-making-billions-from-environmental-destruction.html

UNEP. (2016). *The Rise of Environmental Crime: A growing threat to natural resources peace, development and security. A UNEP-INTERPOL Rapid Response Assessment*. Nairobi, Kenya.

UNU. (2017, January 15). E-waste in East and Southeast Asia jumps 63 percent in five years. United Nations University. Retrieved from https://unu.edu/media-relations/releases/e-waste-in-east-and-south-east-asia-jumps-63-percent-in-five-years.html

Velychko, L. (2015, December 10). Ukraine and toxic chemical cleanup under investigation. Organized Crime and Corruption Reporting Project. Retrieved from http://www.toxicremnantsofwar.info/ukraine-conflict-24-months-of-urgent-environmental-recovery-will-cost-30m/

Weir, D. (2015, July 21). Ukraine conflict: 24 months of urgent environmental recovery of costs $30 million. *Toxic Remnants of War Network*. Retrieved from http://www.toxicremnantsofwar.info/ukraine-conflict-24-months-of-urgent-environmental-recovery-will-cost-30m/

Yardley, J. (2014, January 29). A Mafia legacy taints the earth in southern Italy. *The New York Times*, A4.

Chapter 10

Addressing Environmental Issues for the Future

While preceding chapters of this book described environmental crime in terms of features of criminality and control efforts that centered on the past and present, this chapter tackles the subject of what environmental threats and their control may look like in the future. As with any type of future projection, much of the information here can be considered speculative. However, these projections are grounded in what we know presently about trends of pollution expansion, legal mechanisms that are being put into place to control environmental violations, waste treatment, and a change in the presidential administration taking place at the time of the publication of this book. The chapter begins by acknowledging expected changes in the threat of environmental pollution globally and segues into how government, particularly in the United States, is expected to react to regulation and criminal law responsibilities associated with the production and disposal of toxic substances.

The Global Danger of Hazardous Pollutants

Globally, hazardous pollutants are expected to rise rapidly in the future. Certain sectors of the world are seen as being more vulnerable than others. For example, in the United Arab Emirates (UAE), experts have projected that by 2020 the country would be producing approximately 160,000 metric tons of hazardous wastes per year. That figure is based upon an annual projected growth of 9% per year. However, those same experts add that the growth could be more around 10% or 11% per year. This growth of hazardous wastes in the UEA is based upon how the country is diversifying economically into new industries, including metals, minerals, and fertilizers. Add to that the increase in household products like energy-saving light bulbs, electric batteries, and mobile phones. About 43% of the country's hazardous waste comes from Abu Dhabi and another 36% from Dubai. The biggest producer of the wastes is the oil and gas industry. This industry is responsible for the contamination of soil in exploration and drilling areas. Although oil refineries are responsible for treating any water they use before discharging it into the sea, toxic sludge is often dumped into waterways. Wastes in the UAE can legitimately be disposed by mixing the wastes with other chemicals that can render the wastes innocuous and, in other, cases burned in specially equipped incinerators. However, many of the incinerators are over 15 years old and don't always comply with industry standards. Within the seven Emirates, Dubai has the only officially dedicated hazardous waste facility. Although oil and gas companies do have the means of treating waste properly, other industries do not. It has been reported that wastes from paper mills, pharmaceutical companies, and plastic manufacturers are often dumped in ill-equipped waste landfills or disposed into sewer systems. While steps are being taken to improve proper treatment and disposal facilities in the UAE, it is presently unclear how well those steps will progress in the future to keep up with the increasing rise of hazardous waste volume there (Tordorova, 2012).

Looking at the wider picture in the Middle East, it is anticipated that the rise in hazardous waste will continue to be a malignant problem in the future for the Gulf Cooperation Council (GCC). The GCC is an intergovernmental political and economic union that consists of Saudi Arabia, Kuwait, Oman, Qatar, and Bahrain, along with the United Arab Emirates. By the year 2020, it is expected that these countries, as a group, will produce over 120 million tons of hazardous waste per year. The rapid rise of hazardous waste volume in the GCC will necessitate concentrated efforts on "waste-to-value methods" like recycling and "waste-to-energy" operations. The consulting firm Frost and Sullivan has projected that, for the GCC, the market potential for treating hazardous waste can increase anywhere by 1.5 to 2 times by 2022. They stress that there needs to be a greater focus on optimizing the waste segregation process. This would need to be done at source and material recovery facilities to limit the amount of waste diversion to landfills. The hurdle here is that the composition of waste itself for the GCC has been changing quickly recently. In the past, the waste composition has been more related to construction and demolition operations. The projection is that there will be a pronounced increase in electrical and electronic waste and biomedical waste in the near future. Frost and Sullivan actually see opportunities in the future for the emergence of companies that can deliver solutions related to waste segregation, recycling, and the development of new technologies to complement waste-to-energy services. The firm's outlook is that municipalities located in the GCC have the opportunity to speed up optimization through the evaluation of successful treatment models in other regions and applying them to municipalities in the GCC (Global Recycling, 2017).

Around the world, a serious problem to face in the future with regard to the environment is the ongoing threat of increases in air pollution. In February of 2017, an "atmospheric brown cloud" over South Asia was having a great economic impact on Nepal. Primarily, this economic impact affects the tourism industry in that area. In 2017, a 3-kilometer-thick toxic cloud hovered over much of the region. From 2014 through 2017, Nepal experienced a significant drop in tourism, a major source of revenue for Nepal. The atmospheric brown cloud (ABC) was made up of man-made pollutants such as toxic ash, sulfate, nitrates, aerosols, and black carbon. It was projected by the United Nations Environment Programme that such clouds would continue to affect the Indo-Gangetic plain. This is a densely populated area that covers parts of Pakistan, Bangladesh, India, and southern Nepal. Contributing to these clouds is the smoke from slash and burn agriculture, automobile emissions, and industrial pollution. The ongoing practice of farmers in northwestern India of setting tracts of land on fire to get rid of rice paddy stocks has contributed to environmental problems there. Of great concern is that particulate pollution levels have spiked in other areas of Nepal besides the southern region. An example is that black carbon concentrations on Mount Everest have spiked in 2016 to extraordinary levels, marking the first time in decades that the mountains were not clearly visible for multiple days in a row. This spiking was attributed to increases in fine particles emanating from burned biomass and forest fires (Gyawali, 2017). India, particularly, has fallen victim to ever-increasing levels of air pollution. For the first time, in the middle of 2016, India surpassed China in the overall amount of fine particulate matter pollution in the air. Greenpeace had found that fine particular matter pollution levels in New Delhi reached over 120 mcg per cubic meter. This was in comparison to Beijing's 81 mcg per cubic meter. The World Health Organization conducted a study using a simulation model that found that around 570,000 premature deaths in India were caused by exposure to fine particulate matter. Researchers also found that by converting their calculations in 2 years of life lost, the exposure to fine particular matter in India translated to about 3.4 life years lost in a country where the life expectancy is already low, 64 years. Projections are that India will continue to be a nation in which growing air pollution far into the 21st century will take its toll on its populace (Harvey, 2016).

Moving further west across the globe, air pollution has become a consistent problem both for the present and for the near future. In 2017, London was forced to issue its first "very high" pollution warning due to cold, calm weather that stood in the way of the dispersion of toxic emissions from motor vehicles and wood fires. At the time, the Mayor of London, Sadiq Khan, was compelled to issue one of the highest air alerts ever in the history of London to take precautions against polluted air. This alert stemmed from the operations of innovative monitoring systems developed for London to fight against and provide awareness about the effects of air pollution. For the present and the near future, London has suffered from increasingly elevated levels of nitrogen dioxide and carbon particles that have been more commonly associated with the major industrialized cities of Asia. Based upon reports from King's College, it was estimated that the level of nitrogen dioxide in and around London could be reasonably attributed to over 9,000 premature deaths in London each year. London has been no stranger to the dangers of air pollution given its history of dangerous smog in the 1950s that led to the deaths of at least 4,000 citizens and the asphyxiation of high numbers of cattle. Due to the extreme threat posed by air pollution in London, Mayor Khan spearheaded an effort to modernize London's bus fleet such that buses bought after 2018 would be electric or hydrogen powered. Due to the projections of air pollution in other parts of England, other major cities in that country have been turning to London's efforts as a model for the future to fight air pollution. To counteract the effects of increasing air pollution in the future for London, a major push has been underway to introduce "ultra-low emission zones" beyond the year 2020 in which only the cleanest vehicles would be permitted to operate in the zones with all other vehicles forced to pay steep admission charges (Kottasova, 2017).

To effectively consider how countries should deal with the future onslaught of increased air pollution, one need only to acknowledge Iceland's approach to handling the problem of carbon dioxide. In reality, Iceland has not had a severe carbon dioxide problem. This country relies almost exclusively on geothermal heat and hydropower to generate electricity. Consequently, over 80% of all energy used in Iceland is powered through renewable sources. However, Iceland's geothermal energy production is made from volcanic rocks and travels with other gases into steam from turbines at geothermal plants. To address this issue, Iceland started a program called Carbfix. The primary mission of this program is to take carbon dioxide from geothermal plants and store them in rocky ground between 400 to 800 meters in depth. While Iceland's effort is not original in the sense that other countries have stored excess carbon underground, it is original in the sense that it departs from preserving carbon in a gaseous or fluid form. The Icelandic approach for the future is to dissolve carbon in large volumes of water and pump them into porous basaltic rock. This is a volcanic rock that takes form after the cooling of lava. Iceland's approach involves a chemical reaction that transforms carbon dioxide into a carbonate. This carbonate binds with calcium, magnesium, or iron. These elements naturally occur in high amounts in basalt. Early endeavors in this process have demonstrated that not only is injecting carbon dioxide into basalt rock effective, but the carbon dioxide is rapidly mineralized, that is, it is converted into rock. In the course of two years, over 90% of injected carbon dioxide in Iceland has become mineral. According to experts, carbon dioxide can effectively be stored in rock, and that rock stays constant through geological timescales. This process, for the future, can be extremely valuable in that 10% of the rocks that make up continents are basalt, as is most of the ocean floor. Experts have praised this effort for the future and projects like that of Archer Daniels Midland to store large volumes of carbon dioxide in Illinois sandstone layers (Mooney, 2016).

So, the future for hazardous substances is not all bleak. In some countries, the combination of improved treatment techniques, waste minimization, and a lower demand for highly toxic pesticides represents an air of optimism for the future. In Europe, communication and semiconductor technologies, primarily represented by global positioning systems (GPS) and radio frequency identification (RFID), are expected to generate new revenue opportunities for hazardous waste management services. Smart data solutions, such as using routing software to improve cost controls, are expected to be instrumental in improving the efficiency of the collection and treatment of hazardous waste. For Europe, the combination of the introduction of this technology along with enhanced waste source minimization practices are expected to reduce hazardous waste volumes by at least five million metric tons by the year 2021. While Central and Eastern Europe are far behind Western Europe in terms of effective hazardous waste management, it is anticipated that Central and Eastern European countries will begin to close this gap in the future as a result of European Union regulations that stress incentives for effective recycling and recovery of substances. The critical factors here will be full participation in the source segregation of hazardous waste along with the improved tracking of these wastes (PR Newswire, 2017).

The situation in China might not appear too inviting to companies producing highly toxic pesticides, but it should bode well for improving environmental conditions in that country in the future. In 2016, the market situation for pesticides was described as a declining growth rate in revenues for Chinese pesticides manufacturers, a falling market demand, and lower prices of pesticides. In 2016, this led to a depressed market situation even though Chinese producers kept their production high, resulting in an oversupply of these substances. Starting in 2017, technological advancements led to the replacement of many old pesticides, used in China's fields, with more highly efficient and less toxic pesticides. In addition, improvements in crop protection machinery have led to less reliance on the use of pesticides in general. China has implemented a government plan of "zero growth" of pesticides consumption until the year 2020. This move by the Chinese government has enhanced efforts for the development of more highly efficient, low toxic, and "environmental friendly" pesticides. The use of traditional pesticides in China that contained high toxicity and dangerous ingredients will eventually be phased out. Actions such as this represent important new steps for approving environmental conditions in China for the future (SAT News, 2017).

The Outlook for Environmental Crime Control

The publication of this book comes at a time of transitioning from the prior Obama presidential administration to the present Trump presidential administration. President Trump's early statements on the environment during his campaign promised to work to eliminate what he called harmful and unnecessary national environmental policies. These policies included the Climate

Action Plan (a set of strategies intended to guide efforts for climate change mitigation) and the Waters of the U.S. rule (a rule that defines which rivers, streams, lakes, and marshes fall under the jurisdiction of the Environmental Protection Agency (EPA) and the Army Corps of Engineers). Trump was vocally committed to rolling back former President Obama's signature efforts in the area of environmental protection. These efforts included initiatives to fight climate change and to limit fossil fuel production on public lands. On the newly revamped Whitehouse website under the Trump administration, the "America First Energy Plan" was highlighted as the top item in the "issues section" of the site. The site promised to address what were termed "burdensome" regulations on the energy industry. Recognizing that many would fear that environmental protection under a Trump administration would be weakened, it is important to keep in mind that fewer cases had been pursued by the EPA under the Obama administration at the time of the presidential transition than at any time in the prior 20 years. During the final year of President Obama's administration, the number of prosecutions emanating from EPA investigations proved to be half as many as was the earlier part of his presidency or under the Clinton and George W. Bush presidencies. By the time of President Obama's departure, the number of EPA investigators had plummeted to a ten-year low. Some EPA representatives reported that the decline in the number of agents was as a result of budget cuts. However, the outgoing Director of the EPA's criminal division declared that the reason was that EPA officials were afraid of antagonizing Congressional Republicans with large numbers of environmental crime prosecutions. Experts, like former environmental crimes federal prosecutor Peter Anderson, have declared that while the EPA has focused their efforts on high priority cases like major oil spills, lower priority actions like illegal disposal into sewer systems have garnered far less attention by enforcers. These types of actions can indirectly encourage small companies into believing that their activities would fall under the enforcement radar (Murkami, 2017).

An early sign that worried environmentalists under Donald Trump's administration was manifested in remarks made by Myron Ebell, the advisor who led Trump's EPA transition team. In an interview, Ebell stated that the EPA's environmental research reports and data would not be removed from its website but that climate education material could be withdrawn. Backing away from Trump's campaign pledge that he would eliminate the EPA altogether, Ebell talked about "incremental demolition" instead rather than using the route of an executive order. The climate action plan that was created by Ebell for the Trump administration dismisses climate science and proposes a withdrawal from the Paris climate agreement (an agreement within the United Nations Framework Convention on Climate Change (UNFCCC) dealing with greenhouse gas emissions mitigation). It is apparent that any attempt to eliminate the EPA would be orchestrated by Trump's appointee to head the EPA, Scott Pruitt, the former Oklahoma Attorney General who had sued the EPA more than a dozen times in the past (Nelsen, 2017). Key Democrats in the Senate strongly opposed Pruitt's appointment, citing concerns with his ties to the oil and gas industry. In addition, he was criticized for not totally accepting climate change by contending that climate change scientists continue to disagree about whether or not it is real, despite a near consensus on its reality by scientists. In an unusual lobbying effort against Pruitt, the EPA employee union encouraged their members to contact Senators to vote against him. Ultimately, Pruitt was confirmed to head the EPA (Merina and Marsh, 2017).

When Scott Pruitt finally assumed his post in February 2017, he immediately went on the offensive. In his first speech to the EPA, he described a "toxic environment," not referring to environmental pollution but using the term to describe the political rhetoric of his critics. Part of that criticism stemmed from news reports from media sources like the *New York Times* that referred to secret alliances that were purportedly forged by Pruitt with Republican attorneys general and oil and gas companies to undermine environmental regulations. In his speech, he characterized the term "regulation" as meaning something that would make "things regular" and he promised to seek advice from representatives of private firms before imposing new regulations. The topics of climate change and harm to the environment were not included in the speech (Figure 10.1) (Kaufman, 2017).

A worrisome sign for environmentalists came on March 9, 2017, when the EPA Administrator Pruitt publicly announced his belief that carbon monoxide was indisputably *not* a primary contributor to global warming. This public statement solidified the Trump administration position that was clearly at odds with the established scientific consensus on climate change. Pruitt's emphatic statements firmly contradicted years of research and analysis by international science institutions and federal agencies like the EPA. At the time, his remarks served as a signal that the Trump administration would not be satisfied with simply rolling back previously established climate change policies, but would take an aggressive stance on the underlying legal and scientific basis of those policies. These statements were made in the face of conclusions arrived at by the National Oceanic and Atmospheric Administration and NASA that the Earth's average surface temperature rose close to 2.0°F since the end of the 19th century. The two organizations concluded that this rise was primarily due to increased carbon dioxide and other human-made emissions into the atmosphere. The position taken by Pruitt flew in the face of thousands of experiments and observations since the 19th century that demonstrated that carbon monoxide traps heat in the Earth's surface. Environmentalists were further outraged by the fact that some executives of the nation's

Figure 10.1 Former EPA Administrator Scott Pruitt. https://archive.epa.gov/epa/sites/production/files/2017-02/2017.02.21-pruitt-green-room-event_057.jpg

largest fossil fuel producers commented that even they were surprised by Pruitt's comments. At the time of the publication of this book, there was speculation that Mr. Pruitt's comments could lead to placing the Trump administration in violation of federal law. The Clean Air Act defines carbon dioxide as a pollutant that harms human health. Under this Act, these pollutants must be regulated by the EPA. Consequently, the EPA is legally obligated to regulate carbon dioxide (Figure 10.2) (Gillis and Krauss, 2017).

Mr. Pruitt's tenure as the head of the EPA did not last long. A series of ethics investigations regarding unorthodox travel management and his questionable 24-hours-a-day security detail (among other indiscretions) proved to be his undoing, leading to his resignation on July 5, 2018. On February 27, 2019, the Senate confirmed former Deputy Administrator Andrew R. Wheeler to take Pruitt's place as the Administrator of the EPA (Figure 10.3). Critics lamented the appointment, noting that prior to his appointment Wheeler was an active coal lobbyist in Washington and would be prone to repealing environmental regulations.

So, what do these turn of events portend for the future? Some writers like Brad Plumer speculate that there are five possible "alternative futures" for the EPA under President Donald J. Trump. Plumer acknowledges that, for a long time, Donald Trump had spoken about reining in the EPA. But, there are certain hurdles that could reasonably stand in the way of doing this effectively. One possible "future" is that the agency is abolished altogether. Indeed, a House representative from Florida introduced a bill on February 3, 2017, to "terminate" the agency, citing that the agency had a burdensome annual budget of $8 billion and employed more than 15,000 people. Attached to this "future" is that the United States currently has dozens of wide-ranging environmental laws enacted since 1970 that require the federal government to contain air and water pollution.

Figure 10.2 EPA Administrator Scott Pruitt joins coal miners at the Harvey Mine in Sycamore, Pennsylvania, to kick off EPA's *Back-to-Basics* agenda, which was intended to "refocus EPA on protecting the environment, promoting economic and job growth, and returning power to the states." www.epa.gov/sites/production/files/2017-04/33633313030_e7aceb0c30_o.jpg

Figure 10.3 Fifteenth Administrator of the EPA, Andrew Wheeler. www.epa.gov/aboutepa/epas-administrator

To reach this end of EPA termination, Congress would have to either transfer the responsibilities of the EPA to another agency or strive to repeal all of the environmental laws that are presently in place. Consequently, this "future" would seem highly unlikely. A second possible "future" is that Congress decides that the EPA should retreat from addressing climate change. The Supreme Court decided in 2007 that the EPA possessed the authority to regulate greenhouse gases under the Clean Air Act if it was decided that such gases would present a serious threat to public health. In the Obama administration, this decision became the basis for enacting stricter fuel economy standards for cars and trucks, restrictions on methane leaks from oil and gas drilling, and limitations on CO_2 from power plants. In the House of Representatives, 110 Republicans signed a bill that would eliminate all authority over greenhouse gases that the EPA currently has. The likelihood is that Senate Democrats would effectively filibuster this bill for at least two years, stalling it considerably (Plumer, 2017).

The third "alternate future" is that Trump's pick for EPA Administrator will fundamentally tear up all of Obama's environmental policies. Although the Administrator is confined to working within the parameters of the Clean Air Act, he does possess the option of starting a multi-year process of rewriting the Clean Power Plan, a plan that is instrumental in regulating carbon dioxide from power plants. Obama's plan was to promote cleaner energy and cut emissions 30% below 2005 levels by the year 2030. Wheeler may conceivably take a more modest stance and simply encourage coal plant operators to improve efficiency. If he desires, Pruitt could also attempt to rewrite EPA's "Waters of the U.S." rule to limit the Clean Water Act to a limited number of streams and wetlands. Expectations are that Pruitt has a greater chance of achieving this future than other "alternate futures." Another possible "future" for the new presidential administration is that Congress will neutralize the EPA through carefully calculated budget cuts. This alternative future may carry some weight in that it is expected that the Republican-controlled Congress will try to reduce the budget for the EPA and add riders to limit the EPA's authority over regulation. The final "alternative future" for the environment in the United States is that the EPA will emerge from the Trump years fully intact. Opposition to efforts to weaken the authority of the EPA may prove to be extremely strong. It is conceivable that this opposition can lead to lawsuits from opponents and, consequently, if these lawsuits succeed, the EPA may not experience much transformation at all. One can harken back to how President Ronald Reagan and President George W. Bush attempted to downsize the EPA during their administrations. Some of their efforts did weaken certain aspects of the EPA. But other efforts were stymied in court. An example of this was during the Bush administration when states sued the EPA and were able to get the U.S. Supreme Court to instruct the agency to effectively deal with greenhouse gases. Only time will tell how any of these alternative futures play out for the future of the EPA and environmental protection in general in the first half of the 21st century (Plumer, 2017).

Discussion

This chapter has offered some projections for what may be in store for environmental protection in the future. The chapter began by describing how present environmental threats still exist in many parts of the world and, in some cases, may be exacerbated in the future due to special circumstances. The chapter pointed out that the growth of new industries in certain parts of the world, like the United Arab Emirates, is leading to more perplexing problems for effective waste treatment for the future. It is clear that other countries in the Middle East are encountering great challenges to effectively address rises in hazardous waste, especially in the areas of electronic waste and biowaste. These challenges will remain in the future unless rigorous efforts toward enhanced waste segregation and containment are implemented. An area of great urgency for the future is the increase in air pollution, particularly in countries like India where life expectancy has been reduced because of exposure to air pollution. While all these factors endanger human health, there is reason to believe that sensible responses to these threats can serve to help minimize their effects. Cities like London have engaged in special initiatives like the development of ultra-low emission zones and government support for increases in electric and hydrogen-powered vehicles. In other parts of Western Europe, communication and semiconductor technologies should continue to make treatment more efficient in the future. While Eastern and Central Europe have fallen somewhat behind in waste treatment, a new emphasis on recycling and other environmentally conscious practices should hopefully bring these regions up to speed with Western Europe. And, in China, the technological exploration of developing less toxic pesticides has lessened the threat of unhealthy pesticide exposure in that country.

The chapter pays a good deal of attention to what the future may hold for the United States considering the change in the presidential administration and the early signs that the administration would deemphasize the importance of environmental protection nationally. Given statements made by the new president as well as statements and actions by the new EPA Administrator, it appears that the direction that the new administration is headed in does not bode well for strong environmental protection. Under the Trump administration, global warming is still questioned, and a softening of regulations appears to be imminent. While the elimination of the EPA has been mentioned, it is not believed to be a likelihood. But a weakening of the EPA is a strong possibility for the future. Environmentalists caution that this may open the doors for an increase in regulatory and criminal violations against the environment. In any case, the shifting struggle between "pro-environment" forces and "pro-business" forces in the United States should continue on a national scale for years to come.

Discussion Questions

1) This chapter reflects upon trends taking place presently involving the continued generation of harmful pollutants in different parts of the world. In some regions, the production of pollutants continues to increase with no visible attempt to contain them. In other regions, the government and the private sector have endeavored to confront growing pollution through innovative responses. Considering this, some may see the future environmental situation as a glass "half empty" while others may see it as "half full." How do you see it? Are you optimistic about how the world will respond to environmental threats in the future, or are you pessimistic? Explain your answer thoroughly.
2) Over the years, the EPA has consistently reduced the number of investigators they employ, and the number of criminal prosecutions emanating from EPA investigations has steadily declined. A position held by many is that a weakened EPA would be an invitation for those who would contemplate committing environmental crimes to commit more crimes. An opposing position is that environmental violations could be handled better by state governments than by the federal government in that more power should be given to local and state governments to handle environmental protection. How do you feel about this? Provide a strong justification for your answer.

References

Gillis, J., & Kraus, C. (2017, March 10). Chief of EPA bucks studies about climate. *The New York Times*, 1.

Global Recycling. (2017). *Future opportunities in the Gulf region: Sustainability will be the governing force, driving the change on how waste is addressed in this region.* Retrieved from http://global-recycling.info/archives/905

Gyawali, S. (2017, February 8). Air pollution problem could disrupt Nepal's tourism industry. *Earth Island Journal*. Retrieved from http://www.earthisland.org/journal/index.php/elist/eListRead/air_pollution_problem_co uld_disrupt_nepals_tourism_industry/

Harvey, C. (2016, May 11). Air pollution in India is so bad that it kills half 1 million people every year. *The Washington Post*. Retrieved from https://www.washingtonpost.com/news/energy-environment/wp/2016/05 /11/air-pollution-in-india-is-so-bad-that-it-kills-half-a-million-people-every-year/?utm_term=.8fa8bf248bf8

Kaufman, A. (2017, February 21). Scott Pruitt goes after critics, his own staff in first speech to EPA. *Huffington Post*. Retrieved from http://www.huffingtonpost.com/entry/pruitt-epa-speech_us_58ac7e76e4b0c4d51057164f

Kottasova, I. (2017, January 24). Here's what London is doing about its pollution problem: The problem is deadly serious. *CNN Money*. Retrieved from http://money.cnn.com/2017/01/24/news/economy/london-air-poll ution-alert/

Maharashtra, M. (2016, April 10). GCC waste management industry to present untapped opportunities, notes frost and Sullivan. Retrieved from https://ww2.frost.com/news/press-releases/gcc-waste-management-industry-present-untapped-opportunities-notes-frost-sullivan/

Merina, D., & Marsh, R. (2017, January 18). Trump's EPA pick: Human impact on climate change needs more debate. *CNN Politics*. Retrieved from http://www.cnn.com/2017/01/18/politics/scott-pruitt-epa-hearing/

Mooney, C. (2016, June 9). This Iceland plant just turned carbon dioxide into solid rock – and they did it superfast. *The Washington Post*. Retrieved from https://www.washingtonpost.com/news/energy-environment/wp/2016/06/09/scientists-in-iceland-have-a-solution-to-our-carbon-dioxide-problem-turn-it-into-stone/?utm_term=.863cabc9732c

Murakami, K. (2017, January 2). With fewer agents, EPA cuts back on cases. The Dailey. *Citizen*. Retrieved from http://www.daltondailycitizen.com/news/local_news/with-fewer-agents-epa-cuts-back-on-cases/article_76a9e858-810f-5f04-ad5e-017bdfa5c3bf.html

Plumer, B. (2017, February 13). 5 Possible futures for the EPA under Trump. *Vox*. Retrieved from http://www.vox.com/energy-and-environment/2017/2/13/14533134/possible-future-epa-trump

PR Newswire. (2017, January 11). Tech advancements expand growth opportunities within European hazardous waste management. Retrieved from http://www.prnewswire.com/news-releases/tech-advancements-expand-growth-opportunities-within-european-hazardous-waste-management-300389279.html

SAT Press Releases. (2017, February 10). *China's pesticides market: CAC Shanghai 2017 impacts and forecast.* Retrieved from http://www.cnchemicals.com/Press/89119-China%E2%80%99s%20pesticides%20market:%20CAC%20Shanghai%202017%20impacts%20and%20forecast%20.html

Todorova, V. (2012, September 24). Fears of rapid rise in hazardous wastes. *The National*. Retrieved from http://www.thenational.ae/news/uae-news/environment/fears-of-rapid-rise-in-hazardous-wastes

Index

Pages with *italics* mark photographs.

A Deadly Business (movie) 50
A to Z Chemical Company 12–13
Achenbach, Dr. Laurie 27
acid rain 4
Act to Prevent Pollution from Ships (APPS) 71
active mining 65–66, 107–109, 111, *see also* metal mining
Adler, Freda 60–61, 63, 65, 73–74
Africa 102–103, 112–113
agricultural pollution 20, 107, *see also* inorganic pesticides; pesticides
air dispersal 73
air pollution 2, 4, *16*, 32–33, 73, 95, 116–117
Alabama 44, 94
Alaska 8, *70*
ancient Rome 2–3
Anderson, Peter 119
Angel, Peter 42
Animas River 67
Argentina 107
Arizona 48
Arthur Kill 11–13
asbestos 62–64, 74
ATP Oil & Gas Corp 72–73

bacteria 27–30
Bazalgette, Joseph 4
Beatrice Foods, Inc 7
Belize 107
benzene 48
BHP Billiton 108
bioremediation, case study 28–30
Bolivia 107, 111
Bonn Convention of Migratory Species 112
Boxer, Barbara 49
Brazil 78, 108

Brimblecombe, P. 2–3
Brown, Lester, *Hazardous Waste in America* 11
Browner, Carol 93
Brownstein, R. 5
Bryant, Bunyan 90
Bullard, Robert 90, 94–95
Bureau of Safety and Environmental Enforcement (BSEE) 83
Burns, Robert 96–97
Bush, George W. 123
byproducts 15, *see also* hazardous waste

California 47–48; Department of Toxic Substances Control 48; Environmental Protection Agency (CalEPA) 84–85; Mecca 48–49
Canada Steamship Lines 77
car oil changes 45
carbon dioxide 117–118
carbon monoxide 120
Carracino, William 11–12
case studies: environmental justice 96–99; fracking 86–87; hazardous waste neutralization 28–30; illegal hazardous waste disposal 11–13; New Jersey Organized Crime 55–58; responsible corporate officer doctrine 40–42; ship breaking 75–78; Tunisia 112–113
cathode ray tubes (CRTs) 9–10, 44
Chatham County landfill 97
Chavis, Dr. Benjamin 90
Chemical Control Corporation (CCC) 6–7, 11–13, *see also* entrepreneurial environmental crime
chemical industry, and production-related waste 17–19
chemical manufacturing, and hazardous waste 4, 20–21
Chemical Waste Management 94

chemists, and illegal hazardous waste disposal 51
Chemtronics waste site 28–30
Chevron CPV facility *66*
Chevron Questa Mine 66
Chile 107
China 3, 10, 102, 116–118
cholera 3–4
Ciba-Geigy 7, 51–52, *52*
Ciner Gemi Acente Isletni Sanayi Ve Ticaret S.A 72
civil actions 83
civil administrative proceedings 83
Civil Rights Act (1964) 91
Clean Air Act 5, 32–33, 37, 62, 73, 95, 121–122
Clean Power Plan 122
Clean Water Act 5, 33–34, 37, 72, 87, 123
Climate Action Plan 118–119
climate change 107, 119–120, 122
Clinton, Bill 92
coal 3–4, 10
Coates, Dr. John 27
Cohen, Shire 104
Collington, Michael 12
Colombia 107
Commission for Racial Justice of the United Church of Christ (UCC) 92
Comprehensive Environmental Response, Compensation, and Liability Act (CERCLA) 7, 35, *see also* Superfund
concentrated animal feeding operations (CAFOs) 20
conspiracy 38–39, 63
containment buildings 23
contaminated soil, and non-point pollution 20
corporations, and criminal liability 39
Costco, illegal dumping by 44–45
cradle-to-grave tracking systems 47–48
Cressey, Donald 46–47, 49, 55, 60
crimes 54–55; calculating 46–47, 49; and opportunity 47–49; pressure 46–47; rationalizing 49–52, 55, *see also* environmental crimes
Cuyahoga River 5
CVS, illegal dumping by 45

D'Allessandro, K. 40–41
Dangerous Ground (Alder) 73–74, 106
DDT 4, 10, 91
deep well injection 22
Deepwater Horizon disaster 8, 72
Department of Commerce, and the Endangered Species Act 36
Department of the Interior 82–83; and the Endangered Species Act 36

DeVroom, Dawn 46, 52; *Is Illegal Dumping of Hazardous Waste a Viable Business Strategy?* 44–45
Djerba 112–113
Dotterweich, United States v 39
DSD shipping 72
Duane Marine Salvage Corporation 12–13
Duke Energy 8, *9*

E-waste 10, 103, 109–110; cathode ray tubes (CRTs) 9–10, 44; increasing problems of 8–9; and Third World countries 9
earthquakes, from fracking 69
East/South-East Asia 109–110
Ebell, Myron 119
Ecology Center 16
EkoTek Inc. 41
El Salvador 107
electrical utilities sector 19; pollution reduction efforts 19
elephants 102
Elizabeth River 6–7
embezzlement 49
Emergency Planning and Community Right to Know Act (EPCRA) 8
Endangered Species Act 36
Eni Oil 104
entrepreneurial environmental crime 46, *see also* Chemical Control Corporation (CCC); Northeastern Pollution Control
environmental crimes 54–56, 111; and environmental justice 95; in Italy 104; *knowing* actions 37–38, 40, 42–43; and organized crime 56–58, *see also* crimes
Environmental Defense Fund 82
environmental equality 89, 91
environmental equity 91
environmental justice 89, 93, 95–99
environmental justice movement 91–93
environmental offenders 60, 75, *see also* routine offenders/routine environmental crimes
environmental racism 89–90
environmental violation categories 61; land dispersal 61–65; mining-generated land and water dispersal 65–69; water dispersal (ocean-going vessels) 69, *70*, 71–72
EnviroSolutions 58
Epstein, Samuel, *Hazardous Waste in America* 11
European Commission 76–78
European Maritime Safety Agency 78
European Union 76–78, 118
Executive Order 12,898 (3 C.F.R. §859 [1995]) 92–93, 95–96
Exxon Valdez 8

Fagin, Dan, *Toms River: A Story of Science and Salvation* 51
fair treatment, defined 93
false statements 38
Federal Insecticide, Fungicide, and Rodenticide Act (FIFRA) 36
Federal laws/regulations 32–36, 39, 42–43
Fish and Wildlife Service 83–84
Florida: Department of Environmental Protection (FDEP) 85–86, 99; Fish and Wildlife Conservation Commission (FWC) 85; Fort Myers 86, 94, 98–99
forestry crimes 102–103, 108
fossil fuels: and lead pollution 4; and plastics 16, *see also* fracking; gasoline; petroleum; petroleum industry
fracking 10, 68–69, 86–87
fraud 38

General Accounting Office (GAO) 91, 94–95
General Electric Company 73
Genovese Crime Family 56
Gold King Mine, Colorado 67, *68*
Greenpeace 82
groundwater, protection of 25–26, 30
groundwater monitoring wells 25
Guatemala 107
Gulf Cooperation Council (GCC) 116

hazardous waste: and chemical manufacturing 4; defined 1, 15; health impacts of 35, 44, 62, 104, 109–110, 117; lack of understanding of 52–54; sources 15, 17, 20–21, 56, *see also* byproducts
Hazardous Waste in America (Epstein, Brown, and Pope) 11
hazardous waste sites, locations of 91–92, 95
Hong Kong Convention for the Safe and Environmentally Sound Recycling of Ships 77
Hooker Chemical Company 6–7
Hopkins, Jack 42
houbara bustard 112
Hunt, James 97–98
Hunt, Max 28
hydraulic fracturing processes. *see* fracking

Iceland 117–118
ignorance, deliberate 53–54
illegal drilling operations *70*
illegal fisheries/fishing 101, 103, 113
illegal hazardous waste disposal 11–13, 22–24, 41–42, 44, 55, 107, 113; asbestos 62–63; employee compliance 50, 69–71; and ignorance 53–54; in New Jersey 11–13, 42, 55, 57–58, 61–62; in New York 73, *see also* crimes

illegal wildlife trade 101–102, 106, 110–112
"in situ remediation" 28–30
incineration: and hazardous waste disposal 22, 24–25, 28; molten-salt method 27
incompatible wastes 26; improper storage of 64; storage of 23–24
India 7–8, 30, 116–117
Indiana, West Calumet Housing Complex *65*
industrial/boiler furnaces 24–25
Industrial Revolution 4, 10
injection wells 29, 68
inorganic pesticides 4, 10, 91
inspectors, detecting crimes 47
Interagency Task Force on Electronics Stewardship 9
International Labor Organization (ILO) 76
Interpol 101
iron, bacterial treatment of 27
Iron Oxide 13
Is Illegal Dumping of Hazardous Waste a Viable Business Strategy? (DeVroom) 44–45
isosaccharinic acid (ISA) 27
Italy 104–106, 111–112

Jerusalem, ancient 3
Johnson & Towers, United States v 42

Kaufman, Harold 50
Kerkennah archipelago 112–113
Keystone Policy Center 68
Kin-Buc waste disposal facility 11–12

Lake Erhai 3
land dispersal 61–65
landfills 97; and environmental racism 90–91; and groundwater 25–26; and hazardous waste disposal 22, 25–26; incinerator ash 25; plastics 17; wind dispersal 26
Latin America 107–108, 111
Lavorgna, Anita 106–107
law enforcement officers (former), and illegal hazardous waste disposal 51
law enforcement officials: Adler on 60–61; conspiring with offenders 63, 74
lead 3, 9, 48
lead pollution, Ancient Rome 2, 10
legal hazardous waste disposal: costs of 22–23; deep well injection 22–23; incineration 22, 24–25, 28; landfills 22, 25–26; restrictions on 27–28; treatment processes 22, 26–28
Legambiente 104
lime sludge 98–99
London (U.K.) 3–4, 117, 123

Los Angeles Times 47–48
Love Canal 6–7

M/T Bow Lind 69–71
M/V Artvin 72
M/V Neameh 69
M/V Thetis 69, 71
M/V Trident Navigator 69, 71
Macaluso, Charles 50
McDonald & Watson Waste Oil Company 40
MacDonald & Watson Waste Oil Corporation, United States v 40–41
Madonna, Steven J. 55–58
Maersk Group 77
Mafia 104–105
Mahoning River 68–69
Makra, L. 2–3
Massari, Monica 105–106
Master Chemical Company 40–41
meaningful involvement, defined 93–94
mercury 107
metal mining 18, 20, 66, 108, *see also* active mining; mining abandonment; primary metals
metal pollution 3
methyl ethyl ketone 48
methylmercury 107
Michigan: Civil Rights Commission 96; Flint 96, 99
midnight dumping 47, 69, 104–105
Miller, Steven 41
mining abandonment 65–68, 74–75
mining-generated land and water dispersal 65–69
Model Penal Code 39
Mohammed, Arifet 112
molten-salt combustion 27, *see also* incineration; treatment processes
Monzini, Paola 105–106
Myanmar, illegal wildlife trade 102

Narragansett Improvement Co. (NIC) 40
National Air Toxics Program 62
National Ambient Air Quality Standards (NAAQS) 32
National Emission Standards for Hazardous Air Pollutants (NESHAP) 32–33
National Oceanic and Atmospheric Administration 120
National Pollutant Discharge Elimination System (NPDES) permit program 33
National Strategy for Electronics Stewardship (2011 report) 9
navigable waters 33–34
negligence 37

Nelson, and Bill 99
Nepal 116
neutralization processes 26
New Jersey 6–7, 11–13, 51, *52*, 55–58, 61–62; A-901 58; Department of Environmental Protection (DEP) 58
New Mexico 66
New Source Performance Standards (NSPS) 33
New York 6–7, 62–64; Department of Environmental Conservation (DEC) 73, 84; Kensington Towers Complex 63, *64*
Newark Bay 11–13
NGO Shipbreaking Platform 76
Nightingale, Florence 3
Nixon, Richard 81
Noble Drilling LLC 69–70
Noble Drilling Unit Kulluck *70*
non-governmental organizations 82
non-point source pollution 19–20
North Carolina 8, *9*, 28–30, 89; Buncombe County 28–30; Department of Transportation (NCDOT) 97; Warren County 89–91, 96–98
Northeast Hazardous Waste Research Project (Rebovich) 45–46, 49–52
Northeast Pollution Control 12
Norway 77
nuclear reactors 108
nuclear waste 27, *see also* radioactive wastes

Obama, Barack 119
ocean-going vessels 69, *70*, 71–72, 75–78
Oceanfleet Shipping Limited (Oceanfleet Shipping) 71
Oceanic Illsabe Limited (Oceanic Illsabe) 71
offshore oil platforms 72–73, 75
Ohio 68
oil spills 8, 30, 45, 107
oil tankers 8
Organization for Economic Co-operation and Development (OECD) 77
organized crime: and e-waste 103; and environmental crimes 56–58; Mafia 104–105

pangolins 102
Paris climate Agreement 119
Park, United States v 39–40
Pellow, David 91
Pennsylvania 4, 10; Clean Streams Law 86; Department of Environmental Protection (PADEP) 86–87
pesticides 20, 36, 118, *see also* inorganic pesticides
petroleum 4

petroleum industry 19, 21
plastics 15–16, 17
Plumer, Brad 121
polychlorinated biphenyls (PCBs) 4, 10, 96
Pope, Carl, *Hazardous Waste in America* 11
post-World War II period 4–5, 10, 15, 22, 66
President's Council on Environmental Quality 91
pressure, and environmental crimes 46–47
primary metals: and production-related waste 17–18, *see also* metal mining
protests 97–98
Pruitt, Scott 119, *120–121*, 123
Puerto Rico 107
Pure Food and Drug Act 39–40

questions: environmental crimes 55, 78; environmental justice 99; EPA 124; federal regulations 43; global environmental crimes 111–112, 124; historical hazardous waste 10–11; industrial hazardous waste 21; legal hazardous waste disposal 28; state-level enforcement 87

Racketeer Influenced and Corrupt Organizations (RICO) Act 38
radioactive wastes 6, 11, 25, 27, 113, *see also* nuclear waste
Rebovich, D. 23–24, 45
Refuse Act. *see* Rivers and Harbors Act (1989)
Rege, Aunshul 106–107
Regional E-waste Monitor: East and Southeast Asia 110
responsible corporate officer doctrine 39–42
rhinos 102
rivers, polluted 4
Rivers and Harbors Act (1899) 34
rotary kilns 24–25, 28, 73
routine offenders/routine environmental crimes 45–46
Royal Caribbean Cruise lines 46
Royal Dutch Boskalis 77
runoff 20, 26
Russia 108, 111

Safe Drinking Water Act 34
Salmane, Mohamed 112
Samarco 108
Schlichtmann, Jan 7
Self, Steven 41–42
sewage treatment plants (STPs) 5
ship breaking 75–78
Ship Recycling Regulation 76–77
Sierra Club 82
situational environmental crime 45
Slade, Frances 40

small waste generators 47, 49–50, 55
Snow, Dr. John 3
soil/sand, in Italy 106–107
solid waste collection, and organized crime 57–58
Solving the E-waste Problem (StEP) Initiative 103
source reduction 18–19, 21, 26
stabilization processes 26
State Implementation Plans (SIPs) 32
statistics: E-waste 109–110; illegal fishing 103; illegal wildlife trade 102; improper disposal of hazardous waste 56; minorities and hazardous waste sites 94; oceangoing vessels and monetary penalties 69; projected pollution growth 115–116; ship breaking 76
storage tanks, for hazardous waste 23
strict liability 37–38
Superfund 7, 35, 68, *see also* Comprehensive Environmental Response Compensation and Liability Act (CERCLA)
surface impoundments 23
Swannanoa, North Carolina 28–30
Switzerland, and Ciba-Gegy 51
synthetic organic chemicals 4

T/V Green Sky 72
tanning factories 3
Target, illegal dumping by 45
Tennessee, asbestos-related crimes 63
The News-Press 98–99
Third World countries: and E-waste 9; ship breaking in 78
Tiber River 2
Toms River: A Story of Science and Salvation (Fagin) 51
Toms River Chemical Corporation 51, *52*
Toxic Substances Control Act (TSCA) 35–36, 95–96
Transpetro 78
treatment processes: bacterial 27–28; bioremediation case study 28–30; and hazardous waste disposal 22, 26–27
treatment/storage/disposal (TSD) facilities 23, 56–57, 73, 75; and illegal disposal 23–24, 56–58, 62, 106
trichloroethylene (TCE) 29
Trump, Donald 118–119, 121, 123–124

Ukraine 108–109, 111
uncontrolled hazardous waste sites 94
Union Carbide/Bhopal (India) 7–8
United Arab Emirates 115–116, 123
United Kingdom 3–4, 4, 117, 123
United Nations Environmental Programme (UNEP) 101, 107, 111, 116
United Nations Framework Convention on Climate Change (UNFCCC) 119

United States: air pollution in 4; Alabama 44, 94; Arizona 48; Arthur Kill 11–13; California 47–49, 84–85; Ciba-Geigy 7; Colorado 67, *68*; Cuyahoga River 5; Deepwater Horizon disaster 8; Elizabeth River 6–7; Florida 85–86, 98–99; Indiana *65*; legal hazardous waste disposal 22; Love Canal 6–7; New Jersey 6–7, 11–13, 51, *52*, 55–58, 61–62; New Mexico 66; New York 6–7, 62–63, *64*, 84; Newark Bay 11–13; North Carolina 8, *9*, 28–30, 89; Ohio 68; Pennsylvania 4, 10; Tennessee 63; Times Beach, Missouri 7; Utah 41; West Virginia 86; Woburn, Massachusetts 7

United States Department of Justice (USDOJ) 83–84

United States v Dotterweich 39

United States v Johnson & Towers, Inc 42

United States v MacDonald & Watson Waste Oil Corporation 40–41

United States v Park 39–40

uranium mines 66, *67*

U.S. Air Pollution Control Act (1955) 4

U.S. Army Corps of Engineers 13, 34, 86, 119

U.S. Coast Guard 46, 69–72

U.S. Commission on Civil Rights 94

U.S. Environmental Protection Agency (EPA) 2, 5–8, 19, 32–33, 34, 42–43, 65, 81–83, 93, 95, 119–124; and asbestos 62; and California 49; and Chevron 66; Criminal Investigation Division 83; F-list 2; K-list 2; landfill minimum standards 26; Office of Research and Development 81; Toxic Release Inventory 16–17; and the Toxic Substances Control Act (TSCA) 35–36; TRI National Analysis 17; and uranium mines 66, *67*

U.S. Federal Water Pollution Control Act (1972). *see* Clean Water Act

U.S. Food and Drug Administration 12

U.S. Geological Survey 69

U.S. Oil Pollution Act (1990) 5, 8

U.S. Resource Conservation and Recovery Act (RCRA)(1976) 1–2, 6–7, 9, 25, 34–35, 41–42, 49

U.S. Supreme Court building *33*

Utah 41

Venezuela 107

Wal-Mart, illegal dumping by 44–45

Walgreens, illegal dumping by 44–45

Ward, Robert 96–97

Ward Transformer Company, Inc. 96

waste containment methods 23

waste oil, from ocean-going vessels 69

waste piles 23

waste rock 66

waste trafficking 103–107, 109, 111

wastewater: from fracking 68; from illegal drilling operations 70

wastewater dumping 10, 68–69, 108

water dispersal: ocean-going vessels 69, *70*, 71–72, 75; offshore oil platforms 72–73, 75

water pollution 19, 67–68, 76, 95

Waters of the U.S. rule 119, 123

West Calumet Housing Complex *65*

West Virginia, Department of Environmental Protection (WVDEP) 86

Wheeler, Andrew R. 121, *122*, 123

Wholesale, illegal dumping by 45

Wolf, S. 5

W.R. Grace & Co 7

XTO Energy 86–87

Yunnan (China) 3, 10

Zimbabwe 102